BEYOND THEIR LIMITS OF LONGING

Contemporary Writers & Veterans on the Lingering Stories of WWI

Edited by
Jennifer Orth-Veillon, PhD

Foreword by
Monique Brouillet Seefried, PhD

MILSPEAK BOOKS
An imprint of MilSpeak Foundation, Inc

Manufactured in the United States of America
Library of Congress Cataloging-in-Publication Data
Orth-Veillon, Jennifer, editor
Library of Congress Control Number: 2022935877
ISBN 979-8-9857941-0-6 (paperback)
ISBN 979-8-9857941-1-3 (epub)

Cover Photo: James Grieve (https://www.jamesgrievephotography.com/), commissioned by 14-18 NOW, in partnership with the National Trust, Activate Performing Arts, Creative Foundation, Eden Project, National Theatre Scotland, Nerve Centre, Sunderland Culture and Taliesin. In association with the Aberystwyth Arts Centre, Magna Vitae, MOSTYN, Swansea University and Swansea Council, SeaChange Arts, Theatre Orchard, The Grand Theatre of Lemmings, and Visit Blackpool. Sand portraits designed by Sand in your Eye. Supported by the National Lottery through the Heritage Lottery Fund, Arts Council England, and Big Lottery Fund, and the Department for Digital, Culture, Media and Sport. With additional support from Backstage Trust, Bloomberg Philanthropies, Calouste Gulbenkian Foundation (UK Branch) and National Rail.

Cover design by: www.BoldBookCovers.com
Formatting by: Michelle Bradford Art
Editing by: Margaret MacInnis

MilSpeak Foundation, Inc.
5097 York Martin Road
Liberty, NC 27298
www.MilSpeakFoundation.org

*In memory of all servicemembers and civilians
from WWI and from all wars, past and present.*

CONTENTS

2 PAST AND PRESENT: BRIDGING THE WWI MILITARY-CIVILIAN DIVIDE

FOREWORD

From 2017-2019, Jennifer Orth-Veillon rendered an immeasurable service to the world of literature and knowledge of WWI by creating and producing "The WWrite Blog: Exploring WWI's Influence on Contemporary Writing and Scholarship," which serves as the inspiration for this collection. Throughout the WWI Centennial, she brought back to light the work of writers who wrote during and after WWI about the conflict, and most importantly, gave them a new voice online through the writings of veterans of other wars and scholars who revisited the experiences of soldiers, medical personnel, and civilians. *Beyond Their Limits of Longing: Contemporary Writers and Veterans on the Lingering Stories of WWI* reunites the blog posts in a dynamic series of essays offering diverse aspects and points of view about the period, from combatants to non-combatants, to minorities and women, to friends or foes—all present enriching perspectives on the past that speak to today's world issues more than ever.

At a time when the world is suffering through a pandemic never seen since the Great Influenza of 1918, and when unending military conflicts devastate entire regions of the Middle East and Africa, while millions of refugees are trying to escape war zones or poverty, the essays and creative pieces published in this collection shed light not only on specific events such as the U.S. withdrawal from Afghanistan or the Covid-19 crisis, but also on experiences and emotions pertaining to all aspects of the human condition. Animals and plants are not forgotten here and the devastating effects of warfare even they endured call upon us urgently to take care of our planet's future to reduce the negative impacts our human actions have on its well-being.

Raised with a passion for archaeology, and after spending an important part of my life dedicated to the study of history and Near-Eastern archaeology, I am well aware that, for centuries, history was written by the victors, and has described the world through the eyes of rulers, military leaders, and the educated class. As an archaeologist, I have always been mindful of a sentence from the ancient *Egyptian Book of the Dead* stating, "To speak the name of the dead is to make them live again."

When I think of the monuments and memorials to WWI dead all over France, my country of origin, with their long list of soldiers' names, many from the same family, I wonder how to make their memory live again. Yes, these lists of names, these works of art are meant to ensure that they are not forgotten. They indeed remind us of the sacrifices made by so many nations around the world at the time of WWI, but do they give life again to those who lived, fought, and died during WWI? So few give us a glimpse of the soldier's experience. They rarely describe or immerse us into the feelings, the emotions, the suffering not only of those who fought but also of those waiting for them at home, of those whose life was totally disrupted or even destroyed by the war.

This collection does all this, and so much more. Each and every piece invites us to another journey, another approach to the world we live in today, through war narratives connecting the present to WWI literature and the cultural, historical and military narratives of the past century, its conflicts and its unending search for peace. Jennifer Orth-Veillon's blog curation for the Centennial not only served the mission of the WWI Commission but also offered many contemporary writers a common platform to lead us into new ways of thinking about WWI and what it means to us. I commend her initiative and hope this book will perpetuate these memories to captivate and educate many more generations of readers.

Monique Seefried was born a French citizen in Tunisia and became a U.S. citizen in 1985. She holds a PhD in history from Sorbonne University in Paris. She is fluent in English, French, German, and Italian. She spent the last forty years working in education and the museum world. A curator for twenty years at the Carlos Museum of Emory University, she became in 1999 the founding executive director of CASIE (Center for the Advancement and Study of International Education). From 2003 until 2009, she chaired the Board of Governors of the International Baccalaureate (IB). President of the Croix Rouge Farm Memorial Foundation, she was appointed in 2014 a commissioner on the U.S. WWI Centennial Commission and will serve until the inauguration in 2024 of the WWI National Memorial in Washington. A knight in the French Order of the Academic Palms, in the Order of Merit and in the Order of the Legion of Honor, she is also an officer in the French Order of Arts and Letters. In 2019, she was awarded the Secretary of Defense Medal for Outstanding Public Service.

INTRODUCTION

JENNIFER A. ORTH-VEILLON, PhD, editor

> *They were mortal, but they were unconquerable.*
>
> > -Willa Cather from *One of Ours (1922)*, inscription on WWI
> > Memorial at Pershing Park in Washington DC

ZEITGEISTS

In an interview on the eve of the WWI Centennial in November 2018, I asked Pulitzer Prize winning writer and Vietnam War veteran Robert Olen Butler why he set his Christopher Marlowe Cobb detective series during 1914-1919. He cited comparisons between what he called that era's *zeitgeist* and the big issues of today. From the WWI-period, he listed:

> There were drastic new technologies, expanding the capabilities of mass killing and destruction. There was America's ascendance to a premiere place in the world. There were war-driven waves of immigration, often desperate in motive; the struggle for a viable free press; violent acts of terror; the thrashing of governments under siege; the clash of ideologies–both political and religious. There was racial oppression and gender repression. There were dictators and would-be-dictators gaining and asserting power.

He concluded by asking, "Does all that sound familiar to you?"

Deadly drone technology, the U.S. withdrawal from Afghanistan, migrants drowning by the thousands in the Mediterranean, social media on trial, the January 6th storming of the Capitol, China's emergence as a possible superpower, the Russian invasion of Ukraine, MeToo, the Black Lives Matter provide but a few parallels. I interviewed Butler in 2018, about two years before the Covid-19 outbreak, which might have allowed him to draw a further comparison between today's issues and WWI by citing the Spanish flu global pandemic of 1918.

To answer Butler's question—yes, WWI, undoubtedly, is also the story of our times.

Beyond Their Limits of Longing: Contemporary Writers and Veterans on the Lingering Stories of WWI was not conceived to suggest a slide backward into time, but, as Butler suggested, to use WWI to offer insight and lend historical context to contemporary global crises. By asking some of the most well-voiced writers on war today to generate creative and critical perspectives on WWI, the aim of this collection is to broaden the reader's understanding of ways in which history and literature foster cultural platforms for problem solving and, ultimately, peacemaking.

FROM BLOG TO BOOK

Beyond Their Limits of Longing: Contemporary Writers and Veterans on the Lingering Stories of WWI is informed and inspired by The WWrite Blog. Exploring WWI's Influence on Contemporary Writing and Scholarship, a blog I curated for the Official United States World War One Centennial Commission's website from 2016-2019. The blog featured posts almost every week by emerging and established writers from all genres who volunteered to reflect upon the place of WWI memory in the United States and in the world.

I was motivated to create a book because, though a blog is easily accessible in some many respects, its contents will eventually be removed and housed by the National Archives, where they will be less visible. In addition, the blog is loosely categorized and grouped by subject matter or themes. Therefore, it is difficult to grasp the detailed breadth of the content without clicking through some 120 posts one by one. I believe that, while blockbuster films like Sam Mendes' *1917* ignited momentary public interest in WWI, much of the website material, unless showcased in book format, would be left unread soon after the 2018 Centennial. It is my hope that by transforming selections from the blog into the print and digital collection, *Beyond Their Limits of Longing: Contemporary Writers and Veterans on the Lingering Stories of WWI,* more organized access will be granted to teachers, students, scholars, veterans, and the public who wish to examine and explore the contemporary and dynamic impact of WWI on war writing.

FROM FRANCE TO THE U.S.

A move to France in 2015 provided the impetus for the blog. Far from the iconic WWI battlefields of the northeast, the important presence of WWI was nonetheless obvious while driving through the countryside and cities in the Beaujolais region. In the center of even the least-populated village stands a monument commemorating the WWI dead and, often, the loss is the greatest in the smaller places. Corsica, one of the least densely populated areas in France and geographically the furthest away from the Front, incurred more losses than any other region. While France commemorates veterans from other wars in parades and ceremonies on November 11th, the national holiday, still called *l'Armistice 1918*, remains focused on WWI.

Yet this is what I had expected. Compared to WWII and Vietnam, WWI seems like a forgotten war in the U.S. In classes on war literature I taught at the university, most of my students were unaware of its impact on our country. Before the new WWI memorial was built in Pershing Park, national memorials commemorating all major wars except for WWI had their place in Washington D.C. Compared to France, the U.S. lost far fewer lives and fought most of it on foreign soil. France provided the main stage for WWI's Western Front and battles like Verdun destroyed large parts of the rural landscape.

However, what seemed to me at first a large public French narrative about WWI when I moved to France, 1914-1918 also proved to be neglected subject. I taught in French universities and was surprised by my students' lack of interest or knowledge in the war that had killed approximately 1,700,000 citizens, about 4.3 percent of the population. Near the entrance of one private school, a wall commemorating the 150 students and teachers killed went unnoticed. Amid the memorials, the parades, and the days off, a current discussion of the Great War and the lasting effects on France was missing. It appeared that the most celebrated icons of war–the individual soldiers, the *poilus,* and their individual experiences—had not outlived the monuments.

Frustration over the lack of individual voices in public war narratives was a theme I found not only in the WWI and other war literature I studied with my students while teaching at the Georgia Institute of Technology but also in the creative writing workshops I led for student-veterans of Iraq and Afghanistan. In each case, soldier-writers I encountered from past or present wars expressed resentment because the soldier's experience had been neglected by other kinds of stories told by the media, Hollywood, or history books. When Gulf War veteran Seth Brady Tucker and Iraq War veteran Brian Castner spoke to my class about their writing, both specifically expressed that this phenomenon of neglect—and the way veterans write about it—was first articulated by WWI writers like Wilfred Owen and Ernest Hemingway. Tucker even claimed he learned to both read and write poetry while studying Wilfred Owen in a desert foxhole.

I realized today's war writers owed a debt not only to the service of WWI soldiers but also to the unprecedented way they wrote about the war. WWI was one of the first times in history that literature played a large role in unveiling a side of war little shown to the reading public. Returning servicemembers did express a traditional renewed sense of patriotism and pride following the Allied victory. The violence of battle was romanticized, and stories of bravery or sacrifice glorified. However, a new wave of writing gave way to a vision of war that illuminated absurdity, horror, and the crippling long-lasting physical and psychological consequences of life afterward. Hemingway called the glorification of war after his service in Italy "obscene." Nurse Ellen La Motte described combat injuries as grotesque and pathetic.

In today's world of 24-hour media streams, terms like these convey unsurprising descriptions for ubiquitous mass violence. However, by writing about the less-heroic aspects of war, writers like La Motte and Hemingway did something revolutionary. The strong soldier fought bravely and didn't complain. The weak soldier was a coward and a criminal. While patriotism, triumph, and heroic sacrifice are certainly important aspects of the combat experience, they do not paint a complete portrait of the long-lasting effects of war on soldiers, on families, and on the community. It could be said that WWI writing, for the first time in history, was responsible for exposing the severity, variety, and complexity of war wounds to the public.

Why WWI? It, as Butler mentioned in his interview, inevitably had to do with the unprecedented elements this war introduced to an unsuspecting world—the unbreakable nationalistic alliances formed by powerful empires, the misery of inch-by-inch trench warfare, masses of soldiers suffering deep psychological damage ("shell shock"), new weapon technology that disfigured the human body beyond recognition and razed entire cities in seconds, entire populations wiped out not only by war, but also by the Spanish flu epidemic that swept the continents. In combat, Russia, France, the British Empire, Germany, and Austria lost close to a million soldiers each and their wounded nearly doubled that number. America officially entered only in 1917 but lost around 53,000 soldiers in combat during just seven months in 1918.

The Vietnam War serves as an interesting point of comparison—this conflict lasted fourteen years and the combat dead totaled around 47,000. Covid-19 provides another point of comparison. Since the first confirmed case was reported in January 2020, over 900,000 people have died in the U.S. In addition to the combat deaths, the Spanish flu epidemic, considered to the worst pandemic in modern history before Covid, cost Americans another approximately 675,000 lives between 1918-1920.

Still, what emerged immediately from WWI's destructive aftermath was a dominant public narrative of triumph and commemoration. Many of the WWI literary works considered today as classics offered a counter-narrative. Some were even censured. Voices of African American soldiers and women fell into silence. At the time I began the blog, the WWI Centennial Commission had no substantive information about WWI literature although I found such sites elsewhere. Yet what I did not find was information about how WWI had continued to shape literature, writing, scholarship, and art in today's world—an approach that would encompass a diversity of traditional and non-traditional WWI voices: African Americans, women, Native Americans, the enemy.

A HYBRID COLLECTION IN TEN THEMATIC SECTIONS

Beyond Their Limits of Longing: Contemporary Writers and Veterans on the Lingering Stories of WWI is organized in ten thematic sections. It aims to appeal to a diverse audience and is therefore a hybrid collection of personal essays, non-fiction pieces, scholarly work, poems, and fiction.

While veterans are prominent throughout this collection in other sections, those featured in Part 1 "WWI and Veteran Voices of Today," explicitly draw lines connecting their experience as veterans today to the writerly lives and minds of American soldiers from WWI. "The war to end all wars" ended over 100 years ago, but America has placed troops in global conflicts throughout almost every decade since 1918 and many call the Global War on Terror, which began in 2001, the "forever war." As in WWI, today's veterans come home to further struggles as they may face difficulties reintegrating civilian duties, adjusting to family life, or reassessing the value of their service. Staggering statistics show the prevalence of crime and suicide. Many also go on to lead satisfying lives and become accomplished leaders in a variety of fields. To give voice to the range of issues, veterans from Vietnam, Lebanon, Bosnia, the USS *Cole* Bombing, Iraq, and Afghanistan use literature to reflect upon the ways that WWI has shaped their perspectives not only military conflicts around the world but also on their role as American citizens.

Part 2 "Past and Present: Bridging the WWI Military-Civilian Divide" flips perspectives, featuring essays by civilians who discuss WWI as inspiration for their work on American veterans. One of the biggest challenges faced by soldiers coming home from WWI or any war is confronting the disconnect between the civilian understanding of war as represented by the media and the reality of war as experienced by those who witnessed it firsthand. The scholars and writers in this section suggest that learning about WWI and its soldiers can help bridge the civilian-military divide in today's world. David Chrisinger uses letters from WWI soldiers to teach writing to contemporary veterans in Wisconsin and Philip Metres discusses his poetic projects that reflect his father's work as a psychotherapist to veterans. He also takes on the experience of his uncle, a WWI veteran, who died, unbeknownst to the family in a VA hospital. Mark Whalan traces the U.S. government's efforts to reintegrate wounded WWI veterans into civilian life; and Stéphanie Trouillard, a journalist, writes about her centennial Twitter campaign.

Part 3 "WWI and Women Too: Fighters, Nurses, Writers," mirrors the Me Too movement that has stirred women from around the world to both stand up and to fortify their fight for physical, emotional, and intellectual equality. The Centennial provided the opportunity to reveal the forgotten role that women played in WWI and to demonstrate that the war could not have been fought—or won—without them. A common misconception is that women only served as nurses, but the reality is that during that time, the Navy and Marines accepted 13,000 women into active duty. Many of the over 22,000 American women who served as nurses during WWI, braved grisly scenes of combat to save soldiers' lives. In addition, thousands of women—both military and civilian—have written about their experience during wartime, leaving precious eyewitness accounts of America both at the home front and on the battlefield. These women have inspired contemporary women scholars and writers to explore the war through their own research and art.

Part 4 "WWI Mattered for Black Lives" reflects the call for a real reckoning with racist and colonial pasts, knocking down statues, rectifying laws, and bringing the overlooked accomplishments of African Americans to light. The Centennial offered the opportunity for America to learn both about the vital roles African American women and men played in WWI and about the ways that these actions were suppressed by history. Well over four million men of color were mobilized into the European and American armies. Considerable contributions from African Americans helped win WWI for the Allies. Unfortunately, their efforts went unrecognized for almost a century as most of the U.S. refused to acknowledge their service. It is estimated that almost 40,000 African American veterans were denied medals of valor during the war, but only two have been posthumously awarded the military's highest recognition. This chapter is devoted to unearthing information about their participation, their daily lives, and the way they were treated when they went home.

Part 5 "Bravery and Resilience: Native Americans in WWI," echoes the actions of American sports teams who have reconsidered their titles deemed offensive by Native Americans. General John Pershing, commander of the American Expeditionary Forces on the Western Front, wrote of Native American soldiers in 1920, "The North American Indian took his place beside every other American in offering his life in the great cause, where as a splendid soldier, he fought with the courage and valor of his ancestors." Estimates report that over 12,000 Native Americans served in WWI and most were volunteers. Many of these believed that their participation would grant them equal rights. Sadly, they did not receive the fair treatment they expected despite their bravery. This section features two projects that have worked to shed light on Native Americans, their contribution to WWI, and their fight for recognition: Alan Leventhal's search to uncover the service of the Ohlone tribe and Chag Lowry's graphic novel that tells the story of his family's past.

Part 6 "From the Other Side of No Man's Land" counters the idea that we live in a world made up of increasing protective bubbles. Social media, gated communities, and polarizing politics have made us associate more with people like us despite globalization's promise to bring different parts of the world closer together. The Covid-19 crisis has found its way onto all seven continents, but we still struggle to understand that the disease has the power to affect everyone on earth, no matter the geographical location or socio-economic status. As Albert Camus suggests in his 1947 novel about a deadly pandemic that hits the Algerian town of Oran, *The Plague*, compassion for others can be learned by stretching our critical thinking skills and, most importantly, our imaginations. Over 100 years after the end of WWI, the writers in this section evoke this reconciliatory process by focusing on the ways that foes may be seen as friends. The Christmas Truce of 1914 that Anna Rindfleisch discusses is the most well-known event of camaraderie during the war. Mark Facknitz delves into the little-known Tsingtau through the photographs of his grandfather, a German WWI soldier.

Part 7 "Peace Brokered and the Aftermath" turns to conflicts that have origins in WWI but deepened after the Treaty of Versailles and continued into the modern world: WWII, the Cold War, and the Israeli-Palestinian conflict. Dutch historian Peter de Bourgraaf and American novelist David Gilham bring new perspectives to the role the end of WWI played in shaping the Holocaust, including the arrest and deportation of Anne Frank's family in 1944. Simone Zelitch addresses the ramifications of the Balfour Treaty, a statement made by the British in 1917 announcing support for a national Jewish homeland. Veteran Adrian Bonenberger highlights the lost importance of the Brest-Litovsk Treaty, a pact that was fundamental to founding modern-day Ukraine.

Part 8 "WWI Literature: Critical and Personal Reflections," explores WWI's shocks to the individual and collective conscious that Virginia Woolf captured in writing her essay "The Leaning Tower" in 1940: "Then suddenly, like a chasm in a smooth road, the war came." Just as the unprecedented violence of WWI changed warfare forever, poets and writers of WWI manifested a deep, indelible shift in style, form, and content. War was presented to a large reading public to the first time as problematic—politically, ethically, and physically. The influence was so great that some like Faulkner, as Panthea Reid details in a tragic story about her father, felt forced to lie about overseas services and wounds to gain this attention. Critical and personal responses to this literature analyze the immediate and lingering effects of this conflict on style and content. Rachel Kambury calls

The Lord of the Rings an explicit WWI story, and M.C. Armstrong links John Dos Passos to the role of media in today's wars.

While important historical anniversaries often prompt a fresh look at literary and other writings from the time, they also inspire present-day writers to write their own creative responses. Part 9 "Poetic Responses to WWI," features poems written during the Centennial. Faleeha Hassan, in a prose poem, challenges authenticity in writing war poetry, and Jane Clarke narrates her experience writing poems to explore a WWI family archive.

In Part 10 "WWI in Fiction Today," Christopher Huang and RJ MacDonald talk about the demands of writing novels that take place in the WWI-era. Benjamin Sonnenberg imagines battles fought by Thomas Neibaur, the first Latter-day saint to win the Medal of Honor, and Andria Williams shares an excerpt from her novel based on the story of a former British soldier who passed for white in WWI.

In America, WWI became overshadowed by WWII and Vietnam, further diluting the voices of poets, novelists, essayists, and scholars who unknowingly set a precedent for all successive war writers who appear in this collection. Those who survived WWI and wrote about it opened the space for readers and writers alike to explore the complexity both of war's physical and mental horrors and of its historical significance in today's world in crises. From the vast scenes on the battlefield to the fight at the home front, WWI writing and scholarship in this collection allows us, through contemporary perspectives, to inhabit the mind and body of individuals soldiers, doctors, nurses, civilians, and families.

part one

WWI AND TODAY'S VETERAN EXPERIENCE

"Sunlight in the Tall Grass"
and "Sleeping in the Trenches":
Composing Poetry and Music for the WWI Centennial

BRIAN TURNER, veteran

With jazz bassist and recording engineer Ben Kramer, and writer/musician Roel Vertov, I helped compose a WWI-inspired musical piece, "Sleeping in the Trenches" as part of *Love Songs to the Bombers Flying Overhead*, an album inspired by war experiences and produced by Roel Vertov and the Retro Legion, an international group of musicians and artists. "Sleeping in the Trenches" called upon the dynamic mixing of flugelhorn, upright bass, baritone, string arrangements, and vocals to create a soundscape/landscape for enriching conversations on WWI. Roel Vertov is also an avid ambient sound recordist, and so he recorded and layered in several versions of rain to help create the soundscape to this piece. I sang the vocals, along with Skip Buhler and Major Jackson. Ben Kramer and Roel Vertov performed bass, string, and horn. The WWI Centennial web designer and producer Theo Mayer created a video, which appeared with the blog post. It can be viewed for now on Vimeo.

As a veteran of Bosnia and Herzegovina and Iraq, I hear the reverberation of war in new ways each time I listen and perform it.

While I can't demonstrate "Sleeping in the Trenches" in this essay, I want to share another poem, inspired by WWI, that was also put to music by Roel Vertov and me, "Sunlight in the Tall Grass:"

After the battle,
After the exhausted troops push the lines beyond the far hills,
We're left with these ruined voices in the field,
Cattle grazing among the dead.
The dull light of an afternoon buzzing with flies,
The low drone of honey bees
And that breeze shimmering in the cyprus along the river.
This beautiful and unlivable day.
Sunlight ambers and burns over the blades of grass.

Sunlight turns to gold on the far ridgeline.
Burns its way into the far hills and is gone
And the soldiers in the trenches, do they look back sometimes,
To think of us here,
Those left behind in the tall grass?
The cattle in the field,
The barn on fire,
Here, in this beautiful and unlivable day.

Brian Turner is the author of a memoir, *My Life as a Foreign Country*, and two collections of poetry, *Here, Bullet* and *Phantom Noise*. He's the editor of *The Kiss* and co-edited *The Strangest of Theatres*. He's published work in the *New York Times,* the *Guardian, National Geographic, Harper's,* and other fine journals. He is a Guggenheim Fellow, and he's received a USA Hillcrest Fellowship in Literature, an NEA Literature Fellowship in Poetry, the Amy Lowell Traveling Fellowship, a US-Japan Friendship Commission Fellowship, the Poets' Prize, and a Fellowship from the Lannan Foundation. He lives in Orlando, Florida.

Interview with Robert Olen Butler:
WWI and the Christopher Marlowe Cobb Thriller Series

ROBERT OLEN BUTLER, veteran

JENNIFER ORTH-VEILLON, PhD

WWrite: The Christopher Marlowe Cobb series is set in the WWI-era. The characters are built inside of WWI and WWI is built inside of the characters. Why did you choose this period?

Robert Olen Butler: I've always been a student of history and I was led to this time by a consideration of the big issues, the Zeitgeist of the 1914-1918 era. There were drastic new technologies, expanding the capabilities of mass killing and destruction. There was America's ascendance to a premiere place in the world. There were war-driven waves of immigration, often desperate in motive; the struggle for a viable free press; violent acts of terror; the thrashing of governments under siege; the clash of ideologies—both political and religious. There was racial oppression and gender repression. There were dictators and would-be-dictators gaining and asserting power. Does all that sound familiar to you?

WWrite: Yes, it sounds like reading today's newspaper.

Robert Olen Butler: Yes it does. I wrote the Christopher Marlowe "Kit" Cobb books to tell the story of our times, as well. And if you're paying attention to our times, it will clearly suggest that at the heart of all things—at the heart of the human condition—is chaos. But the artist who writes fiction has a deep instinct that behind the apparent chaos of life on planet earth, there is order and meaning.

WWrite: Don't your Cobb books, in their creation, not only look outward to draw on the events of the world but also inward to your personal life?

Robert Olen Butler: Yes indeed. Cobb, the main character, is covering the war, but he also happens to be a spy for the United States of America. He grew up as the son of a very famous stage actress, a single mother, who gave him some talents and inclinations in that realm. The interesting thing is that Kit Cobb is probably closer to me in terms of my autobiography than almost any character I've written for more than a decade. I was first an investigative reporter, and then the editor-in-chief of a business newspaper. I went to war. I was trained as a Vietnamese linguist and in Army counterintelligence as a special agent. My dad was the chairman of the theater department of St. Louis. I was trained as an actor before anything else. Kit Cobb is deeply rooted in who I am and the historical espionage thrillers of his era are absolutely the right elements that I need to articulate narratively what I see about the human condition.

WWrite: The historical research and your attention to detail is outstanding. I'm thinking about the windows you described during the sinking of the Lusitania in the second book in the Cobb series, *The Star of Istanbul*. What kind of research did you do to grasp the minutiae of each scene?

Robert Olen Butler: I do a lot of initial research about the larger, tectonic shifts in politics and so forth of the time, but when I'm writing, I'm meticulous about the little things, the quotidian details. But the important thing is that I always do research through the eyes, the sensibility, the goals, and the yearnings of the characters. This means I do a lot of research on the fly. It's the moment-to-moment-based flow of experience that vivifies literary fiction.

What was the dress and what were the streets paved with? And what's the make of the taxicab that Kit has stepped into? What's the smell of the street? I could not have written these books properly even ten years ago. Indeed, I feel sometimes I should be dedicating these books to Google, because the word-searchable resources of the millions of off-copyright full-text books are crucial. That includes all the magazines and newspapers of the era where the details of life reside. We're in the midst now of the books of the 1920s opening up to free access. And fortunately, all the books of the period 1914-1918 are off-copyright.

I'll give you an interesting example. You mentioned *The Star of Istanbul* and the Lusitania sinking. Kit's task as his spy is to follow a German agent, but he also becomes interested in a woman who's a famous silent film star. He approaches her first-class cabin. And given the circumstances, there should be a "do not disturb" sign hanging on that door. A question arises the like of which I ask literally hundreds of times when writing a Cobb novel: were there "do not disturb" signs in 1915, and would one have been hanging on a door in a passenger ship? It took me about four minutes on Google Books trying out combinations of search terms. I found a bound volume of *Hotel Monthly*, a trade magazine for hotel owners, all of it digitized. And sure enough, there was an article in there about the use of "do not disturb" signs on hotel doors

WWrite: We have Instagram, Facebook, and other forms of social media that allow us to communicate instantaneously. But back then, unlike us, they read a lot of letters because the mail came several times a day. Did you get a chance to read letters and if so, how did they play into your research?

Robert Olen Butler: I love to go after memoirs and letters because those are rich troves of detailed information. But interestingly, this whole Cobb series began from one other piece of correspondence from that era. For years, I was a collector of old picture postcards. As I say, I've always had this interest in history. The WWI-era sits in the center of the heyday of the picture postcard as a mode of communication. My collection was made up mostly of cards between 1906 and 1918. And, as you mentioned, in almost every city and town in the country and probably the world, there were at least two mail deliveries a day.

This was also a time when automobiles weren't common, and you could live 20 miles away from somebody and that's a long way. That's a big bite of day to get there and back, but you could sit down in the morning and write a postcard and put it in the mail. It would go out that morning. It would arrive at the recipient's house that afternoon. The recipient could write a response the next morning and put it in the mail. Postcards were the early 20th century text messaging. This was also the era of the Brownie camera, so that people could take their own photographs and then have them developed onto postcard stock.

So one of the cards in my collection was just such a "real photo" postcard. The image on the front was a man in a white shirt and dark trousers. We see him from behind. He's walking on a sidewalk along a run of shops that have Spanish signage on the fronts. You see somebody on horseback in the distance, and he looks like he's wearing a military uniform. Then in the mid-distance, there's a little gaggle of women, and approaching them is the walking man, who I take to be the writer of the card. He's drawn an arrow pointing at one of the women. Now, what I've not mentioned so far is that he has just passed two dead bodies. Two Mexican men lying in pools of blood on the sidewalk. The message on the back of the card is very simple. It says: "after the battle."

So I did some research from all this circumstantial evidence and I concluded that the guy in uniform was an American. The postmark was 1914. Three years later, Woodrow Wilson will finally get us into WWI, but in 1914, the United States of America invaded Mexico because we did not like the Mexican general who had become the president. We invaded to protect American oil interests.

From all this I was inspired to write a short story, "The One in White," which was included in my book of fifteen stories, *Had a Good Time*, each one inspired by a postcard from my collection. *The Atlantic* published it, and it was one of the more recognized stories in the collection, winning a National Magazine Award in fiction.

But the voice of the man, who was unnamed in the story, would not let me go. So eight years later he became Christopher Marlowe Cobb. The first book in the Cobb series is called *The Hot Country*. It's about that invasion.

WWrite: What was it about the that character that didn't go away?

Robert Olen Butler: First and foremost is his voice, which everything about the card allowed me to hear. Almost literally his voice. This comes back to my acting training. My best writing is by a voice-driven narrative persona. It's like inhabiting a character the way a Method actor does. Also, I mentioned the word yearning earlier. Yearning is just a deeper-diving word to express the classic Aristotelian "objectives" and "goals" of the central characters in narrative, which Aristotle said then must be challenged to move the story forward. Plot is simply yearning, challenged and thwarted. And I happen to believe that in literary fiction, to borrow a phrase from Einstein, there's a kind of "unified field theory" of yearning. I believe that if you look deeply enough into the central character of any work of literary fiction, they are yearning for a self, for an identity, for a place in the universe. Why is that so? Because it is true of all us in the lives we lead. That's what the human condition is about. We are all of us, always, striving to answer that enduring question: who the hell am I? If you look at all the flashpoints of our culture and our times—race, gender, ethnicity, sexual preference, politics, religion—what are they? They are simply prefabricated answers to that question. But if you seize on one of those to define yourself, you also almost inevitably create an otherness, even an enmity, about those who prefab in the alternate way. But in literary fiction, the yearning for self is explored in all its complexity, its nuance, to reveal the deeper humanity we all share.

In my mind, what drives Kit is that he is his mother's son. He has a complex relationship with her and she is part of him. He's trying to figure out which part to keep for himself and which to let go. Part of him is a war correspondent and part of him is a spy. He finds ways to figure out who he is through the women he's involved with. Kit is restlessly in search of his own identity. That's one of the primary things that just won't allow me to let go of him as a character.

And I was glad to embrace WWI for this series because there were so many different aspects to it. War takes us to the limits of human experience, and our yearnings get tested in ways they don't get tested in any other kind of event in history. It's a perfect setting for Cobb because his personal sense of self is always playing out against the larger challenges of the historical moment. I think of that scene in *The Star of Istanbul* where he undergoes the sinking of the Lusitania. There is, of course, the vast, terrible trauma of it. Then we find him a few days later in London and he's been through a horrible thing, but all he can think about is finding this actress. His yearning to find the actress remains fully intact even though he's come up against one of the most crucial and most catastrophic moments of the war. The power of the personal quest asserts and defines itself.

WWrite: Would you now like to read an excerpt from your latest in the Christopher Marlowe Cobb series *Paris in the Dark?*

Robert Olen Butler: Yes. Just the very first few pages, just to give you a sense of his voice:

> In the dark above Paris in the deep autumn of 1915, there were always the Nieuports flying their patterns like sentries walking a perimeter. The new svelte model 11—called *Bébé* by its pilots—operated above the high-flying Zeppelins poised to drop on them in a column of searchlight if the Zepps got by the guns at the French forts to the east.
>
> On a November night, I sat beneath the Nieuports at a table outside the Café de la Rotonde. The weather had been unpredictable. It snowed last week, but tonight it was almost mild. It might as well have been April. And that hammering of engine pistons up above might as well have been French worker bees going after chestnut blossoms.
>
> My drink was a Bijou—the greenery taste of the chartreuse fitting right in with the bees in the night—and I was surrounded by people I couldn't actually see, just vague shapes and spots of cigarette flame. But I knew who they were, the assorted male denizens of the left bank. Artists and professors; students furloughed for six days from Hell; students furloughed for good by a stump of an arm or an empty pants leg; the old, the infirm, the foreigners.
>
> The conversation—at turns hopeful, fearful, or miffed–had been low, as if the Zepps would hear us. And I'd sat away from them, near the street. I had my own brooding and ranting to do, which I kept to myself.
>
> But now the voice rang clearly in the dark.
>
> "Monsieur. You will like one Bijou more?"
>
> I looked up at the shadow hovering above. He'd spoken in English but wallowing the words in his mouth like the French do. He was old enough to have grandsons in the trenches.
>
> "Thank you," I said. "That is just what I need."

I'd replied in French. My French was pretty good. My actress mother, who took on my education in all subjects, knew French well from playing Racine and Corneille in her two long triumphant tours of the continent in the mid-nineties. And from a beau or two of hers along the way.

Before the waiter moved off, I said, "Henri, isn't it?"

"Yes. That is me," he said. "Have I forgotten you, to my shame?" He was speaking French to me now.

"Ah, no," I said. "I heard someone address you."

"Thank you," he said.

I said, "I always like to know the name of the man who will help me become more or less drunk."

Henri laughed a faintly suppressed laugh.

"I'm Kit Cobb," I said.

"Monsieur Cobb, you are American, yes?"

"Yes I am."

"You are here." He paused. I grew up in the theater. I knew how to hear subtext. Here meaning Paris. Meaning Paris deep into the Great War. His silence said: *Though your countrymen are not.* Then he finished formally, courteously. "I am grateful to you."

"Plenty of us will be here," I said, addressing the thing he left unsaid. "To fight. The day is coming."

He lowered his voice. "There are too many professors."

I shot him a smile, though I doubted he could see it. He knew his clientele here amidst the universities of Paris on the left bank. And he knew our American president.

"I share your distaste," I said. Then, so he knew I knew what he was really saying, I added, "For Professor Wilson."

He chuckled and I could even make out his shrug. "But still your countrymen will come?" He asked.

"Yes."

"I pray it will be in time."

"So do I."

"And you, sir. What do you do in Paris?"

Ah, how to answer that.

I was a reporter. A war correspondent. But hobbled, thanks to Henri's government. And part-time, thanks to mine. I was also a spy."[1]

Robert Olen Butler has published eighteen novels and six volumes of short stories, one of which, *A Good Scent from a Strange Mountain*, won the Pulitzer Prize in Fiction. He has also published a widely influential volume of his lectures on the creative process, *From Where You Dream*. His latest novel is *Late City* (fall 2021), which takes place in the nanosecond of the death of the last living World War I veteran. In 2013 he was the sixteenth annual recipient of the career-spanning F. Scott Fitzgerald Award for Outstanding Achievement in American Literature. He teaches creative writing at Florida State University.

Notes

[1] Robert Olen Butler, *Paris in the Dark* (Christopher Marlowe Cobb Thriller, 4), 1st ed. (New York, NY: Mysterious Press, 2018).

Visions of War and Peace:
WWI Literature and Authority*

PHIL KLAY, veteran

In the familiar literature, WWI comes to us, overwhelmingly, as muck and madness, an orgy of pointless violence, insanity on a civilizational scale as waves of young men were sent off with fine words into a meat grinder. It is blood gargling from "froth-corrupted lungs, obscene as cancer."[1] It is the old battalion, "hanging on the old barbed wire."[2] It is "strange Hells within the minds War made," while an oblivious public back home remains smugly, and wrongly, certain that "chivalry redeems the wars."[3]

In *Company K*, a brilliant and sadly underread novel by William March, who served in WWI with the Marines and was decorated with the Croix de Guerre, the Distinguished Service Cross, and the Navy Cross, a soldier tries a hand at writing an honest condolence letter after the death of a comrade. He writes:

> He died in agony, slowly. You'd never believe that he could live three hours, but he did. He lived three full hours screaming and cursing by turns. He had nothing to hold on to, you see: He had learned long ago that what he had been taught to believe by you, his mother, who loved him, under the meaningless names of honor, courage and patriotism, were all lies...[4]

This, according to much of the literature of WWI, is the bitter truth which the war revealed. Yes, there are the occasional Ernst Jungers, who compared the exchange of hand grenades to a ballet and who counted his generation lucky to have been able to serve in such a great struggle, but the vast majority of the famous literature of the war, from Wilfred Owen to Siegfried Sassoon to Jaroslav Hasek, is a collective ode to futility, waste, horror, and despair.

Is this, then, the truth?

In *The Fighters,* CJ Chiver's superb book about the current American wars, he announces that he rejects senior officer views so that his book may channel

> . . . those who did the bulk of the fighting with the unapologetic belief that the voices of combatants of the lower and middle rank are more valuable, and more likely to be candid and rooted in battlefield experience, than those of the generals and admirals who ordered them to action—and often try to speak for them, too.[5]

But of course, it's not just generals and admirals who try to speak for the common soldier, it's poets and novelists too. Wilfred Owen is very self-consciously trying to speak for his men, a move which, as the poet Tom Sleigh has pointed out, when contrasted with the less ideologically inflected (and, to my mind, artistically superior) work of an enlisted soldier like David Jones can seem "at best...heroic posturing in an anti-heroic guise [a]nd at worst...a form of unconscious class condescension."[6] And many of the great writers to come out of World War I wrote in the emerging modernist tradition which had already, prior to the war, established itself in an iconoclastic relationship with tradition and the old order.

Marinetti had demanded we "demolish museums and libraries" 1909.[7] Ezra Pound had tried to move past the "crust of dead English" by founding Imagism in 1912.[8] Stravinsky and Nijinsky had shocked audiences with *The Rite of Spring* in 1913. The spirit animating that generation of artists was revolutionary, tossing down old idols with reckless abandon, and so we must ask: Were honor, courage and patriotism revealed as lies by the war, or by these artists' commitment to the prevailing current of thought among their tribe? Both, probably.

After all, the writings of those non-artists at war paint a more complex picture. As the historian Jonathan Ebel points out in the introduction to his *Faith in the Fight: Religion and the American Soldier in the Great War:*

> Soldiers' and war workers' writings do not allow the honest historian to write a polemical history or a predictable ideological critique. As much as one may love or hate the idea of war, love or hate the men and women who

plan wars and send others to die, love or hate those who profit financially or politically from war, the voices of soldiers and war workers will provide, at most, equivocal support.[9]

And indeed, through soldiers' and war workers' letters and diaries he finds that, though we now like to tell the story of WWI as one of disillusionment, that was not the case for many Americans in the war, who saw it as an opportunity to practice "Christianity of the sword" by which they could find personal and national redemption for a God-chosen nation—and that that narrative was strengthened rather than weakened post-war.[10]

Walter Poague, who flew seaplanes on anti-submarine missions and who would be killed six days before the Armistice, wrote to his mother, "This is not a terrible war. It is the most wonderful war in the world. It is the war which means the real salvation of the world."[11] Red Cross nurse Elizabeth Walker Black noted the "exaltation about being under fire" which shirkers, "with their flabby souls and sluggish blood living selfish lives" would never get to experience.[12] And Infantry officer Vinton Dearing wrote to his mother of the strangely compelling nature of war:

> You go out into the moonlight and feel the place 'holy and enchanted,' a new world, half mystical, a different moon, more wondrous lights; -then some tremendous 155 goes off and shatters your dream...Life is great and the aims of the war are great. It is when you see into the aims with your inner eyes that you see the bigness of it all.[13]

To which truth do we owe allegiance? If we accept Chivers' democratic notion that it is among the voices of lower and middle-rank combatants where we find the truth of war, then our poetically informed ideas about WWI must get pulled in an uncomfortably militaristic direction.

As a writer who was once a Marine and writes about war, I am skeptical that any kind of final authority exists on which we can rest the supposed perspective of "the troops," whatever that might be, and doubly skeptical that such authority then reverts to artists. After all, the budding fascism of the post-war period can be spotted not simply among those who grimly clung on to a cold and fatalistic nationalism through all the bloody years of the war, but also among far too many disillusioned artists and writers, from Céline to Evola to

Huelsenbeck to Pound. The destruction of the old values can lead in many directions.

Veterans and artists can convey to us the thrilling or horrifying or insane intensity of war experience. They can show us the desperate attempts at sense-making done by humans engulfed in it. They can allow war to revolt us or seduce us. They can lie, for sure, but even more dangerously, they can tell terribly impartial truths.

In the 1930s, Walter Benjamin took it upon himself to review a collection of essays, edited by Ernst Junger, about WWI and the possibilities within German nationalism for those of heroic spirit. As Benjamin, who a decade later would commit suicide while fleeing the Nazis, points out, all the authors "were themselves soldiers in the World War and, dispute what one may, they indisputably proceed from the experience of this war."[14] And yet, despite this, and despite the fact when war broke out in 1914, Benjamin spent his time translating Baudelaire rather than fighting, he confidently, and accurately, condemns not only their work but even their ability to write about war itself. In the essay, titled "Theories of German Fascism," he declares:

> We will not tolerate anyone who speaks of war, yet knows nothing but war. Radical in our own way, we will ask: Where do you come from? And what do you know of peace? Did you ever encounter peace in a child, a tree, an animal, the way you encountered a patrol in the field? And without waiting for you to answer, we can say No! It is not that you would then not be able to celebrate war, more passionately than now; but to celebrate it in the way you do would be impossible.[15]

This, it seems to me, places authority where it ought to reside. Not in the lived experience of war, nor in the penetrating yet isolated truths of individual artistic geniuses, nor in the aggregated opinions of a given mass of veterans, but in a careful reader possessed of a vision of peace.

*Penguin Press has given permission for MilSpeak Books to republish this essay, which first appeared in Klay's book, *Uncertain Ground*, published in 2022.

Phil Klay is an author and a veteran of the U.S. Marine Corps. His first book, *Redeployment*, received the 2014 National Book Award for Fiction and the National Book Critics' Circle John Leonard Prize for best debut work in any genre. His novel *Missionaries* was named one of the *Wall Street Journal's* ten best books of 2020. He has written for the *New York Times,* the *Atlantic,* the *New Yorker,* and other places. He teaches fiction at Fairfield University.

Notes

[1] Wilfred Owen, "Dulce Et Decorum Est" *Poetry Foundation* (Poetry Foundation), accessed January 12, 2022, https://www.poetryfoundation.org/poems/46560/dulce-et-decorum-est.

[2] Max Arthur, *When This Bloody War Is Over: Soldiers' Songs of the First World War* (London: Piatkus, 2008), 68.

[3] Adam Thorpe, "Rereading: Ivor Gurney," *The Guardian* (Guardian News and Media, November 10, 2007), https://www.theguardian.com/books/2007/nov/10/featuresreviews.guardianreview4.

[4] Robert Clem and William March, *Company K* (Tuscaloosa: University of Alabama Press, 2009), 101-102.

[5] CJ Chivers. *The Fighters: Americans in Combat in Afghanistan and Iraq* (New York: Simon & Schuster, 2019), Xviii.

[6] Tom Sleigh, "'To Be Incarnational," Poetry Foundation (Poetry Foundation), accessed January 12, 2022, https://www.poetryfoundation.org/poetrymagazine/articles/70066/to-be-incarnational.

[7] Filippo Tommaso Marinetti, "The Futurist Manifesto," The Futurist Manifesto, accessed January 12, 2022, http://bactra.org/T4PM/futurist-manifesto.html.

[8] Guido Cavalcanti, Ezra Pound, and David Anderson, *Pound's Cavalcanti: An Edition of the Translations, Notes, and Essays* (Princeton, NJ, NJ: Princeton University Press, 2016),193.

[9] Jonathan Ebel. *Faith in the Fight: Religion and the American Soldier in the Great War* (Princeton, New Jersey: Princeton University Press),14.

[10] Ebel, *Faith in the Fight*, 14.

[11] Ebel, 37.

[12] Ebel, 72.

[13] Ebel, 15.

[14] Ernst Junger, Walter Benjamin, and Jerolf Wikoff, "Theories of German Fascism: On the Collection of Essays War and Warrior, Edited by Ernst Junger," *New German Critique*, no. 17 (1979): p. 120, https://doi.org/10.2307/488013.

[15] Junger, Benjamin, and Wikoff, "Theories,"125.

"Dulce et Decorum Est:"
Discovering WWI Poetry in an Iraqi Foxhole

SETH BRADY TUCKER, PhD, veteran

One of the first war poems I ever read, like most people, was "Dulce et Decorum Est" by Wilfred Owen. I read that poem in a foxhole during the First Persian Gulf War, and I didn't have a way to translate the title or the last line, but I guessed, incorrectly, that I knew about the suffering all soldiers must endure. I too recognized the flurry of activity that went with the yell of "gas!" and I knew what nerve agents would do to us, and the word on the street was that the 82nd Airborne would be going head-to-head with the combat-tested Republican Guard in Iraq. So, I thought I knew what it might mean to die as dirty and smelly and lonely (and young), as I was then.

It was also not lost on me that I was a soldier, like Owen, and like him, I also fancied myself a poet. It was like looking in the mirror, for just a brief moment. I remember the mindless connection I felt with Owen—we were both war poets, and like him, I would write poems that would speak over generations to future soldiers. What I did not know then was this: he'd died during WWI, a hero and a poet, but a dead one. I had another mindless reaction when I discovered this: I did not want to be like Wilfred Owen.

So, I started reading Siegfried Sassoon, Mary Borden, E.E. Cummings, Ford Madox Ford, Rainer Maria Rilke, Robert Service, Gertrude Stein, then the WWII poets, all the way up to our contemporaries. Some of these war poets survive, some of them do not. Some are brave in combat or service, some are not. All of them, however, are brave in words. In this, they did not shirk their responsibilities, in this they did not turn away from the horrors, the obscenity, the awful trench warfare that gifted some of them to us as martyrs, returned some of them to us broken and ill-used but willing to speak of war honestly.

Their voices are trapped for us in poems that have lasted a full hundred years and will be there for our ancestors a thousand years from now. As a poet who writes about soldiers and war, I doubt I will get that sort of billing—no one will remember my war but the economists and historians perhaps as they examine the fall of a great empire—but every word I write carries with it, I believe, all the heavy burdens of those like Owen, who were stripped of their right to question the morality of the wars in which they fought. I consider myself lucky that I am allowed this luxury. In my mind, WWI snuffed the life from Owen, a poet who would most likely have become one of the world's most famous poets. Just imagine writing some of the most recognizable poems the world has ever known, over the course of a year in combat, at the age of twenty-four. It's astounding.

So, I write about soldiers and war, and I do this with as much humility as I can muster—my hope is that I can honor those who came before, walking the wide road they have broken for us, while blazing a trail, however narrow, for those who will follow.

Seth Brady Tucker is the executive director of the Longleaf Writers' Conference, senior prose editor for the *Tupelo Quarterly Review,* is originally from Wyoming, and once served as an 82nd Airborne Division paratrooper in the U.S. Army. Seth's second poetry collection, *We Deserve the Gods We Ask For,* won the Eric Hoffer Book Award and his first book, *Mormon Boy,* was a Finalist for the Colorado Book Award. His essays, poetry, and fiction have been published widely in such magazines and journals as *December, Litmag, Poetry Northwest, Verse Daily, Blackbird,* and the *Birmingham Poetry Review.*

They Shall Not Grow Old...and Neither Have We

TERESA FAZIO, veteran

In January 2019, amid headlines of U.S. negotiations with the Taliban and lingering Syrian ISIL strongholds, I escaped internet news for an afternoon of 3D immersion in *They Shall Not Grow Old*, the Peter Jackson film that brings previously unreleased WWI archival footage of British soldiers to life. Jackson's team smoothed the jerky frame rates of century-old film, employed forensic lip-readers to sleuth dialogue, and recorded the slams of breech blocks from antique artillery.

But when I forked over twenty bucks for a ticket, I didn't know that what would strike me even harder than this impressive technical reconstruction was the similarity of the youthful soldiers to my own Marines in Iraq in 2004.

At age twenty-three, I was a communications officer on a base between Ramadi and Fallujah, supervising two platoons of rambunctious troops as they set up telephone, radio and data networks and repaired equipment. They wrestled, played smuggled-in guitars, and once—I still don't know how they managed this—barbecued shish kabab and steaks on an oil-drum grill. Sure, one hundred years have brought improvements to efficiency; for a caffeinated "brew-up" in the trenches, one can now simply pop open canned Rip-Its. Modern troops wear more personal protective equipment: flak jackets, Kevlar helmets, floppy shoulder pads and groin protectors, knee pads, gloves, and steel-toed boots. And though my Marines' dental hygiene was far better than that of Brits born at the turn of the twentieth century, I swore I could spot in the trenches my freckled, insightful sergeant who resembled an overgrown Dennis the Menace; my switchboard tech, a former college violin major; and my red-haired chess whiz corporal, who despite his tattooed forearms, still looked twelve years old. Most of my troops were nineteen, the same minimum age as British World War I enlistees, though a hundred years ago, the movie tells us, that rule was often flouted by recruiters who told would-be privates, "you better go outside and have a birthday."

Between helicopter parents and electronic records, it's harder to fake your age now. But teenagers' stubbornness, energy, and idealism—ideal for military purposes—remain the same. And like boys turned into donkeys on Pinocchio's Pleasure Island, neither we nor WWI's youthful Tommies seemed to have considered the consequences of our respective decisions to chase adventure. Even if I had watched this movie as an eager high-schooler, I still would have joined.

In the movie, when a passel of young men filed behind parading troops and rushed the barracks, excited to put on a uniform, I recalled my own naïve fumbling as a ROTC freshman with a newly issued camouflage pack. Close-order drill looked the same, and I could swear I'd scooped dinner into a canteen cup from a similar jerrycan. Hearing the green, untested troops wonder whether they could shoot or bayonet another human, I could think only of their callow bravado. But they would learn soon—Calais was only about a 30-mile boat ride from Dover. To a New Yorker, it's as if a ferry across Long Island Sound marked the difference between peace and war.

France's trenches looked like the Quigley, the Vietnam-inspired swamp we were required to wade, swim, and crawl through at Officer Candidates School in Quantico, Virginia. Amid the mess, these WWI troops held cigarettes in tense, dirt-streaked mouths, cracking jokes to pass the time. Putrid, freezing mud caused frostbite and gangrene. The lucky suffered amputations. The unlucky sank into the quagmire, never to be recovered. I thought of my desert deployment and counted my blessings.

The footage of sleeping men looked like my wire platoon during breaks in our weeklong pull of fiber optic cable under an airfield: napping whenever possible, sweaty, and grimy, with overcoats thrown over them in the same manner as modern poncho liners. After mortars crashed onto the airfield, like camo-clad Sisyphus, my troops would go out to splice them before we got mortared again. Across a century, I recognized the adage that you never hear the shell that hits you. I had expected *They Shall Not Grow Old* to have a distant Brooklyn-hipster, Edward Gorey vibe, but minute by minute, it remained emotionally resonant.

The film's single trench-raid sequence is a composite envisioned during the battle of the Somme. One soldier says that the hours waiting to go over the top of a trench and into no man's land for a raid on the Germans "were the longest and shortest hours in life." Toward the end of the film, a sequence lingers on men from the Lancashire Fusilliers waiting in a country lane to go over a small rise and into battle. German machine guns await them. Some look serious; some look nonchalant. Many are smoking cigarettes. One, resembling a very young Ewan Macgregor, looks directly at the camera, terrified. They are a hundred yards from where they are aiming to attack. As director Jackson relays at the end, that film was shot in the last thirty minutes of many of those men's lives. On the first day of the battle of the Somme, the British Army lost more than 19,000 soldiers— nearly double the number of U.S. troops killed in Iraq and Afghanistan combined. Whether they were bodies receding into the muck or carried on stretchers or carried from Humvees or counted on a SIPRNet toll, I felt protective of these servicemembers a century apart.

Soldiers who survived WWI returned to joblessness, a demolished sense of purpose, and clueless civilian family members who argued points of fact about the war although they hadn't been there. This was the similarity from the movie I found most chilling and touching. Ninety years later, veteran unemployment peaked at nearly nine percent soon after the 2008 financial crisis, and since only about eight percent of Americans have ever served in the military, troops' sense of purposelessness and isolation can linger long after serving. I don't know the half-life of my own sense of alienation, though it's not as pronounced now as it was in the several years after I returned from Iraq. It may be something that lasts, in one form or another, as long as I do.

At the Basic School, our instructors drummed into us that we were stewards, holding the line and taking care to supply subsequent leaders with trained troops and clean equipment. As this fascinating, horrifying, ultimately edifying movie has shown, we veterans were indeed just placeholders, as those Tommies were before us. *They Shall Not Grow Old* captures soldiers' fleeting effervescence. It reminded me of our losses, and of fellow troops I may never see again.

WWI was supposed to be a war to end all wars. Instead, its human aspect spanned all of them. By 2019, so many of us have not grown old—as a result of combat, yes, but also from post-deployment mental illness, substance abuse and suicide, part of the long tail of war wounds. At some point, we will all die. The wars, however, go on.

Teresa Fazio served in the United States Marine Corps as a communications officer, deploying once to Iraq. Her writing has been published in the *New York Times*, *Rolling Stone*, *Foreign Policy*, *Washington Post*, and *The Nation*, as well as several anthologies. Her memoir *Fidelis*, published by Potomac Books, shows a young woman Marine officer's coming of age through the lens of a deployment relationship, and details the uneasy balance women strike while trading femininity for power in an environment where vulnerability is not only taboo, but potentially lethal.

Ernst Jünger: The Modern War Story

ELLIOT ACKERMAN, veteran

Much of the modern literature of war, which was birthed in the trenches of the First World War, has been arrayed into the following moral arc: a naïve, idealistic youth goes to war; he witnesses the horrors and waste; he returns home haunted, or even destroyed by what he's seen; hence war is evil. Obviously there are many variations on this theme—sardonic novels like *Catch-22*, or narratives that only obliquely reference war such as *The Sun Also Rises*—but much of literature adheres to this basic framework and its moralism. And why shouldn't it? What could be redeeming about the slaughter Carl von Clausewitz references in *On War*? Yet it took the unprecedented bloodletting of the trenches, with their poison gas and futile advances into machine gun fire, to codify this conclusion in art. But is there room for another narrative in literature?

Iconic American writers, such as Tim O'Brien think not. In his seminal book on Vietnam, *The Things They Carried*, he wrote, " . . . you can tell a true war story by its absolute and uncompromising allegiance to obscenity and evil."[1] However, one of the most significant works of literature to emerge from WWI came to a very different conclusion about the nature of war. In the final pages of *Storm of Steel*, its author, German veteran Ernst Jünger writes, "Hardened as scarcely another generation ever was in fire and flame, we could go into life as though from the anvil; into friendship, love, politics, professions, into all that destiny had in store. It is not every generation that is so favoured."[2] Few would characterize the lost generation, which perished in the millions among the muck of trench warfare as favored, but Jünger does.

An infantry officer and Sturmtruppen (storm troop) commander, who fought for the war's entirety from 1914-1918, Jünger was wounded as many as seven times and received the Pour le Mérite, the highest military decoration of the German Empire. Two years after the war, he self-published *Storm of Steel*, a memoir of his time in the trenches, to widespread acclaim and went on to have a rich and varied literary career, producing over fifty books before dying at the age of one-hundred and two. Throughout his life, Jünger's views on war proved controversial. Many criticized him for glorifying war, yet *Storm of Steel* rarely, if ever, veers into the polemical. Jünger's descriptions of battle are straightforward, unsentimental reckonings with violence. If at times he marvels at the spectacular, it is often, frankly, because war is spectacular, and to deny this is not to speak truly of war. Or, as the photographer, Timothy Page, said of war, "you can't take the glamour out of that. It's like trying to take the glamour out of sex, trying to take the glamour out of the Rolling Stones."[3]

Jünger's belief in the redeeming nature of violence, as illustrated in these lines, "We stood with our feet in mud and blood, yet our faces were turned to things of exalted worth,"[4] would be easier to dismiss if they were uttered by a chicken hawk. Elsewhere in the WWI's literature we meet many such figures who exalt war while knowing nothing of it, none of whom is as notorious as Kantorek, the school master in Erich Maria Remarque's seminal *All Quiet on the Western Front,* who encourages his students into the trenches while remaining safely at home. Remarque is critical of Kantorek because he behaved, "in a way that cost [him] nothing."[5] Perhaps this is why Jünger's views, which seem closer to Kantorek than to Remarque's anti-war protagonist Paul Bäumer, cannot be so easily dismissed—Jünger's views cost him a great deal.

With the approaching centennial of the Armistice, revisiting the work of Ernst Jünger takes on a particular import as our fractious world continues to teeter from one crisis to the next. His writings, though out of step with the development of contemporary war literature, are perhaps all the more essential because of their frank, and rare, perspective on the mystical allure of violence. For a nation that, at the time, was entering its seventeenth year of war, staring down that allure might be the best way to escape its clutches.

Elliot Ackerman is the author of the novels *2034, Red Dress in Black and White, Waiting for Eden, Dark at the Crossing,* and *Green on Blue,* as well as the memoir *Places and Names: On War, Revolution and Returning.* His books have been nominated for the National Book Award, the Andrew Carnegie Medal in both fiction and non-fiction, and the Dayton Literary Peace Prize among others. He is both a former White House Fellow and Marine and served five tours of duty in Iraq and Afghanistan, where he received the Silver Star, the Bronze Star for Valor, and the Purple Heart.

Notes

1 Tim O'Brien, *Things They Carried* (Boston, MA: Mariner Books, 2022).

2 Jünger Ernst, *Storm of Steel: From The Diary of a German Storm-Trooper Officer on the Western Front* (New York, NY: Fertig, 1993), 329.

3 Michael Herr, *Dispatches* Internet Archive (New York : Avon Books, January 1, 1977), https://archive.org/details/dispatches00herr_2/page/n5/mode/2up?q=glamour, 275.

4 Junger, *Storm,* 330.

5 Erich Maria Remarque, *All Quiet on the Western Front* (New York, NY: Random House Publishing Group, 1987), 42.

Echoes of Sassoon:
Brian Castner Interviews Matti Friedman

BRIAN CASTNER, veteran
MATTI FRIEDMAN, veteran

When Matti Friedman was seventeen years old, he moved from a typical middle-class neighborhood in Toronto to a kibbutz in Israel, to work as a farmer. A few years later, he was drafted into the Israel Defense Forces, and fought in Southern Lebanon in 1998 and 1999, in a conflict that still has no official name. After his military service ended, he settled in Jerusalem and became a journalist with the Associated Press. It took him fifteen years to write about his war. His memoir of that time, *Pumpkinflowers*, was published in 2016, and chosen as a New York Times Notable Book.

More Canadian soldiers died in the Great War than in any other conflict, and its influence can be felt throughout *Pumpkinflowers*. This puts Friedman at odds with most contemporary American veteran-authors, who often reach to other conflicts for comparison—Vietnam for Iraq, and Korea for Afghanistan, have become typical—when writing about their wars.

Brian Castner, American author, journalist, former Explosive Ordnance Disposal (EOD) officer and veteran of the Iraq War, sat down with Friedman to explore the Canadian view of WWI through the lens of the two different Middle East conflicts.

Brian Castner: *Pumpkinflowers* takes its title from two agrarian images from the kibbutz: the "pumpkin" is the outpost you occupied in southern Lebanon, and "flowers" is the Israeli code-name for soldiers killed in action. The horrible consequences of war, your lost comrades, are hidden in such a peaceful natural symbol. I couldn't help but be reminded of the fields of poppies at Flanders. How did the WWI poets affect your writing of this book?

Matti Friedman: I think the guys who wrote about that war understood that the power of the experience doesn't lie in combat descriptions, which can't really be described but rather in the gap between normal life and nature on one hand, and the war on the other. They give us those poppies first, as in the famous poem—and then dead soldiers speak to us from underneath. Or you'll get a poem starring an oblivious rat working both sides of no-man's land, in Isaac Rosenberg's "Break of Day in the Trenches," and through that character you see how sick the whole thing is. In Siegfried Sassoon's trilogy about the war, he spends almost all the first book just writing about horses, cricket, and bucolic England to set up what happens afterward: the fall. I think they were on the right track. I moved to Israel at seventeen but grew up in Canada, where people wear red poppies on November 11 and recite "In Flanders Fields," which was written by a Canadian, John McCrae. Both of my mother's grandfathers fought in WWI, which still looms very large in the country's collective mind. I think that conflict was much less present for Americans our age growing up in the '80s and '90s, maybe because of fresher memories like Vietnam. Do you remember hearing much about it, and did any of the writing affect your attempts to describe Iraq?

Brian Castner: I think you are right. My family's tradition of military service is extensive—Civil War, World War II, Vietnam—but it skips WWI. I'm not alone in that. America joined the war late, fewer served. Most small towns in America have a Civil War monument, and the Vietnam memorial is famous, but the Great War was obscured. I don't think it's freshness or importance but connection. Our grandfathers were in World War II, our dads and uncles in Vietnam, the Civil War in our textbooks and playing out in the civil rights struggle, but World War I was .

. . where? I grew up in Buffalo, on the Canadian border, watching *Hockey Night in Canada*, and remember asking my dad why all the coaches wore red poppies in the month of November.

That said, my connection to WWI started during my first tour in Iraq, because the security for my EOD unit came from the 3rd Infantry Division. They are known as "The Rock of the Marne," for defending that French river in 1918. And the war we were fighting looked similar in that offensive technology—machine guns and artillery in WWI, IEDs in Iraq—was far outpacing tactics and armor.

I think my writing of Iraq, though, was much more influenced by modern culture: cinematic descriptions, chopped up narrative like a hyper-edited music video. The question in a horror movie is always when to show the monster. In my memoir about Iraq, *The Long Walk,* I show the monster on the first page, and then beat the reader over the head with it for the rest of the book.

I admire what you did in *Pumpkinflowers* because it is far more restrained. The violence has a different impact when it's under tension. As you say, many WWI poems don't start in the trench, they start with something natural, pastoral. Did you plan this from the beginning when writing *Pumpkinflowers*? Or did form follow function, since in the pumpkin, you were stuck in your own trench? Do you feel some affinity for those Great War soldiers, in the military experience and not just in literary treatment?

Matti Friedman: There's an energy and a directness to *The Long Walk* that makes it feel urgent. You came back from Iraq disturbed and things hadn't settled. The reader really feels it. I wrote my book more than a decade after the events, so I thought about them differently.

I was looking for inspiration when I was researching *Pumpkinflowers,* and just wasn't finding it until I read Paul Fussell's *The Great War and Modern Memory* and was introduced in earnest to the world of WWI literature. I dove in and found what I was looking for—these voices that were sensitive but also removed, almost cool. I was surprised at how effective that turned out to be, especially in delivering scenes of violence.

And in a strange way, the wars felt similar. Not in terms of the violence, of course—if you count fatalities, the entire war I'm describing in Lebanon was probably a good morning on the Western Front. But in the way the soldiers were static for years, not conquering land or retreating, just hunkered down, waiting, observing the landscape, wondering what the enemy was up to. In Lebanon, we were in these sandbagged trenches, with machine guns. We called our positions "the line," and we had a dawn alert, just like the soldiers in WWI. The scale and context of the war were obviously different, but I think if you'd dropped a 1917 grunt into our outpost he wouldn't have been confused about what to do.

Brian Castner: I'm struck by how our tracks begin in parallel, and then diverge in the final product. Meaning, you and I are the same age, we both grew up on Lake Ontario, only a hundred miles apart, and then we both fought in Middle Eastern wars that featured roadside bombs and general futility. But your memoir is full of floral restraint, and mine is frenetic and technological. To overgeneralize, yours ended up stereotypically Canadian, and mine similarly American. In your future writing, do you think this Great War perspective will continue to hold? Is it baked into your worldview?

Matti Friedman: I never thought of the style question that way, but it's an interesting point. I think both styles as you described them might be attempts specifically to deal with this experience, the soldier's experience, and aren't necessarily going to work in other writing—unless all we're planning to write is military memoirs, which isn't the case with either of us. Right now, I'm writing a book about spies, and while I still think the reserved tone is effective, in this case the elegiac and floral tone of the Great War writers doesn't really fit. It's important to write a war book if you've seen something like that, but you don't want to get stuck there, either in terms of subject or style.

Matti Friedman is an award-winning journalist and author. He was born in Toronto and lives in Jerusalem. His work has appeared regularly in the *New York Times* Op-Ed Section, the *Atlantic, Tablet,* and elsewhere. Friedman's last book, *Spies of No Country: Secret Lives at the Birth of Israel,* won the 2019 Natan Prize and the Canadian Jewish Book Award for history. *Pumpkinflowers: A Soldier's Story of a Forgotten War* was chosen in 2016 as a NY Times' Notable Book and one of Amazon's 10 best books of the year. His first book, *The Aleppo Codex,* won the 2014 Sami Rohr Prize and the ALA's Sophie Brody Medal.

Brian Castner is an award-winning nonfiction writer and veteran of the Iraq War. He is the bestselling author of *Disappointment River*, *All the Ways We Kill and Die*, and the war memoir *The Long Walk*, which was adapted into an opera and named a *New York Times* Editor's Pick and Amazon Best Book of the Year. His journalism and essays have appeared in the *New York Times*, *WIRED*, *Esquire*, the *Atlantic*, and on *National Public Radio*. He is the co-editor of *The Road Ahead*, a collection of short stories featuring veteran writers, and is a two-time Pulitzer Center grantee. His most recent book is *Stampede*.

British WWI Cemetery Uncovered by U.S. Marines During Iraq War[1]

BENJAMIN BUSCH, veteran

By the end of May 2003, we had been told that the war had transitioned into security and stabilization operations. This was a post-hostilities phase and we were to focus on hearts-and-minds projects. General Mattis had ordered our attention to be on school rehabilitation and that required me to go to the city of Kut, capital of the Wasit Province, for which my unit had become largely responsible. It was the only place that I could purchase electrical wire, paint, concrete, pipe, and plaster.

In a back street, my patrol came across the Kut war cemetery. It had been found covered in several feet of garbage by Marines in Task Force Tarawa during the invasion, and they had cleaned and rededicated it to the British.

Turkish and German troops had laid siege to Kut Al Amara, now simply known as Kut, from December 7, 1915, to April 29, 1916, and the British casualties were astounding. Over 22,000 soldiers had been killed or wounded; 13,000 captured; and here in this small plot submerged from the street by four feet lay the remains of 450 of them. I stepped down into the graveyard and was surprised to see the intricate carving on the headstones. The dead here had not been killed during the siege but rather two years later, when the British and their Arab allies had retaken Kut. I was pleased to see the equitable burial of supporting troops from the British Empire lost in her efforts. Beside infantry and artillery soldiers lay a man with the title,

"FOLLOWER."
F. X. D'SOUZA,
INDIAN LABOUR CORPS
21ST AUGUST 1918

His name appeared below an image of a rifle crossing a shovel, bound with a wreath beneath the crown of England and a banner with the Latin motto *LABOR OMNIA VINCIT* (hard work conquers all). Service was what had brought him from India to Kut, following troops with the toil of carrying ammunition and supplies. I had to wonder where the men lost in the siege had been buried. We had not heard of any other cemeteries and had to guess that they had gone into mass graves somewhere nearby, somewhere just beneath us. This place, full of people tossing trash onto the graves, had been the site of a desperate battle, this soil soaked with blood, this ground heavy with lead...

Gravestone of F. X. D'Souza of the Indian Labour Corps located in British WWI Cemetery, Kut, Iraq. Photo by Benjamin Busch.

"Today is Better than Tomorrow": A British Cemetery Revisited Ten Years After Serving in the Iraq War[2]

BENJAMIN BUSCH, veteran

My photographs have been kept in an order absent of chronology: a blind-folded skull, a pile of boots, an English gravestone, a child waving. I never labeled them, just expected I would remember like everyone does. Seven months of circumstantial evidence. Military mobile exchanges only sold 400-speed film, meant to shoot subjects in lower light, so the reduced resolution is noticeable when the pictures are enlarged, their definition becoming increasingly granular, as if composed of pressed dust. The imperfection of vision is at work, the flickering of lines, the involuntary squint to identify, exactly, what you're seeing, the desert going from vast and static to pulsing and immediate, like memory does.

Kut was familiar but disorienting, the streets swollen with shoppers and thousands of pilgrims. The city made the sounds of collision: sharp honks, truck engines, salesmen calling out, and the constant chirp of police whistles. Goats ate the stiff dead grass in the median. The center of town looked battered.

I stopped into the Iraqi Journalists Syndicate office with my interpreter, Khalil. All the government furniture was foreign-made, wood glossy with thick lacquer chipping at the edges. Everyone smoked. They couldn't believe I came by cab from Baghdad with nothing but an unarmed interpreter. A man shook my hand and smiled. He said they have a proverb: "You enter the house from the door, not by the window." The last American journalist anyone can remember came here three years ago in an armored convoy and only stayed for an hour. The manager said, "There is not enough violence for Kut to be news. You look for stories and the city doesn't exist." Another added, "I would rather not exist than make the news here."

A man named Nassir took us to see a cemetery built by the British for the men lost in the battle for Kut during the First World War, its headstones, like its soldiers, brought from overseas. It was the only place in the region that commemorated the people most responsible for Iraq's creation.

Nassir drove in rushes and jolts, stabbing in between cars, everyone around him doing the same, warning beeps constant, lanes filled like blood vessels, everything somehow pumping through. Traffic intensified as we pushed into the marketplace, the streets channeling a reckless braid of pedestrians, motorcycles, trucks, carts, and cabs.

"It's amazing the roads aren't filled with car accidents," I said. "I can't believe no one gets hit here." Khalil interpreted and Nassir laughed. He had a joyful face.

"Happens," he said, smiling.

He'd been assigned as our official escort for the next eight days in the province and had been cheerful since we arrived. Tomorrow we would hear that he hit a young boy on his way to pick us up, and a cousin was sent with another car to drive us in his place. The child died in the hospital, and we wouldn't see Nassir again as his life became suspended between tribal atonement and court. But today he was happy to be on a journey with the only American in Wasit.

We went to the souk, which was easily three times the size it had been in 2003. Even then the market had been a kind of neutral space, uninterrupted by invasion or regime, the stalls passed from generation to generation. Trade was conducted in dinar notes stamped with Saddam's victorious image or in dollars stamped with our own Founding Fathers. Under Saddam, an Iraqi soldier made the equivalent of $120 a year. By 2004 the Coalition Provisional Authority desperately rehired them at a salary of $400 a month. There was immediate, irreversible inflation. Corruption in the military bloomed as soon as wages did. An Iraqi soldier makes $1,000 a month now.

When I first walked through the market in 2003, the concrete pillars in the central square were covered with xeroxed posters of the people "disappeared" by the regime. A week later, near the Iranian border, we would watch as some of them were exhumed from the site of their execution.

Today we passed through groups of women in black, children emerging and retreating beneath their parade of robes, salesmen with handfuls of cash manning caverns of cloth, sandals, and spice. We arrived in the street where the Kut War Cemetery lay. There was a new fence in front, but the rear wall had been pushed down, exposing the site to the city. The concrete cross still stood, but it looked changed, some graffiti spray-painted on it, the U.S. Marine rededication plaque torn off. Children played on a worn dirt clearing where headstones had been, kicking a ball over the soldiers below, one of them digging on the surface with a little toy shovel. Phragmites reeds had overgrown the rest of the lot. The owner of a nearby shop said it had been used as dump, but the Americans cleaned it, and for a while it had flowers and a guard. The guard went unpaid, he added, and "kids" took the headstones and sold them in the market. The cemetery was a dump again.

"Many groups have come to study the cemetery since 2003. The Americans built a new fence to replace the British one. Then the British built one to replace the American one. They all look for the missing stones, but they find nothing. They ask questions, take pictures, and leave. Just like you." Just like me.

Outside Iraq, Kut is a city remembered only as a military disaster. The British had marched a colonial Indian division up the Tigris toward Baghdad but were stopped by the Turks and their Arab allies at Ctesiphon and sent into a retreat that ended here. For five months they were trapped by siege, unable to resupply or escape, pounded by Turkish artillery, finally surrendering at a loss of 12,000 men. Only 420 have graves here; the rest are still buried somewhere underneath the city. I waded into the dense stand of reeds and found a few toppled headstones still largely intact.

1525 PRIVATE
W. HOSKINS
SOMERSET LIGHT INFANTRY
24TH NOVEMBER 1915, AGE 20
LORD REMEMBER ME
WHEN THOU COMETH
INTO THY KINGDOM

The reeds stood ten feet tall and were plumed with clouds of seeds. I watched them sway above me while crouched by a grave in their shade, a cat passing by using the line of fallen stones as a path. It was my birthday, and I thought of all these soldiers, their entire lives marked by nothing but the day they died and now marked by nothing at all. I emerged coated with dust, my hair downy with silky seeds. Khalil had been pacing the perimeter, calling to me every few minutes as if I were lost in the wilderness.

Five headstones taken from graves in the British WWI cemetery and used to fill potholes in a nearby street. Kut, Iraq. Photo by Benjamin Busch, 2013.

Outside the gate, I noticed some bright fragments tamped into the gutter and recognized the elegant carving of a wreath in marble, the crest of a gravestone sunk into gray drain water. Men watched as I took a photograph. They followed as I went up an alley and found an entire headstone face down in concrete used as a step into a gated courtyard. It showed no sign of damage or wear, dug up unbroken precisely for this purpose and moved a mere fifty feet from its grave. Another photograph and Khalil suggested we hurry. A small crowd was gathering. As we left I found more stacked as a stoop in front of a shop. Then, embedded in the street itself were five stones, carts rattling over them, a carved cross facing up. I kicked aside trash and drew attention from nearby vendors who wondered why I was so interested in the pavement.

In contrast to the forlorn British cemetery, the Ottoman memorial is proudly guarded by an Iraqi Arab paid by the Turkish Embassy. He is the fourth in an unbroken line since his great-grandfather dug the graves. He greeted me with his young son beside him, the next guard. Polished plaques on the gate say TURKISH MARTYRS, 1914–1917, in Turkish on one side and Arabic on the other. The white concrete markers inside bear no names, just the raised Turkish star and crescent, dabbed red with too much paint. These fifty Turkish soldiers and seven commanders are all that can be found, representatives of 10,000 lost here. I asked where the rest are buried and the guard swept his arm around the horizon. Iraqis are not interred in Kut. They are carried to the city-size cemeteries in Najaf and Karbala. Only invaders are buried here. The caretaker showed me the visitor's log, and it is a list of foreigners, mostly Turks. I added my name. On our way back to the car Nassir bought diapers for his new baby. Today is better than tomorrow.

Benjamin Busch is a writer, filmmaker, and illustrator. Following a studio art degree from Vassar College, he served as a Marine Corps infantry and light armored reconnaissance officer, deploying twice to Iraq. He's the writer/director of BRIGHT, and appeared on HBO's *The Wire*, and *Generation Kill*. His essays have arrived in the *New York Times Magazine, Guernica, NPR,* and *Best American Travel* Writing. His poetry has appeared in *North American Review, Prairie Schooner, Five Points,* and *Epiphany* among others. The excerpts appearing in this volume are from "Today Is Better Than Tomorrow" (Harper's), and his memoir, *Dust to Dust* (Ecco).

Notes

[1] Excerpted from Busch's memoir *Dust to Dust.* Original copyright for *Dust to Dust,* by Ecco, an Imprint of HarperCollins Publishers 2012. By Benjamin Busch, ISBN 978-0-06-201484-9.

[2] Original copyright for "Today Is Better Than Tomorrow", VOL. 329, NO 1973, October 2014, *Harper's Magazine*: 2014. Harper's Magazine Foundation. Copyright now held by Benjamin Busch.

Accidental Tourism and War Memorials

ERIC CHANDLER, veteran

I was running along the Bow River in Calgary. I stopped at a memorial next to the water because I don't need much convincing to rest. Made up of six vertical stone slabs, it was engraved with the names of Calgary's war dead from WWI through Afghanistan. Next to the slabs, a piece of steel held the words "We Will Remember Them." The stone closest to the river read:

WE LIVED
FELT DAWN
SAW SUNSET
GLOW

I'm embarrassed to say that I didn't recognize these lines. I'll blame oxygen debt. I got back to my hotel room and looked up the memorial and inscription. The Calgary Soldiers' Memorial was dedicated on April 9, 2011, the anniversary of the Battle of Vimy Ridge, a notable victory accomplished by Canadian forces. The inscription comes from "In Flanders Fields" by John McCrae, one of the best-known poems to come out of the Great War. I must've read the poem before, but I didn't remember the words and I didn't realize the author was Canadian. My run along the river by that memorial highlighted my WWI knowledge gap.

Traveling isn't just part of my job. It is my job. I work for the airlines. Which is great, except for one thing: I miss my town on the shore of Lake Superior. Sleeping in my own bed every night after running or cross-country skiing on my local trails is not part of the deal. I fight homesickness by running during my layovers. For an hour or two, I lope around strange towns and distract myself with accidental tourism. Sometimes, I have a specific destination in mind, but mostly I stumble into surprises.

My connections with WWI seemed minimal at first. I was stationed at Luke Air Force Base, Arizona, named after Lt. Frank Luke, Jr., the "Arizona Balloon Buster." I know who Eddie Rickenbacker was. I think I saw a photo of my mom's paternal grandfather wearing a uniform from that time frame. That just about covers it.

Maybe that's why we have war memorials. It places a tangible object in our midst so that we can remember things that happened long ago. These memorials try to give some meaning to the loss and sacrifice. I travel from town to town and happen to run by many memorials. When I'm in Washington, DC, my company puts me up in the hotel where Reagan was shot. I run downhill from there toward the National Mall and swing past the World War II memorial to the Korean War Veterans Memorial, the nineteen statues wearing ghostly ponchos, frozen in time. To the Vietnam Veterans Memorial, where I slow to a respectful walk, per the directions of the sign. I walk down the gentle slope to the vertex of black stone, all the names are somewhat underground. Now I know that the Vietnam Veterans Memorial is the former location of the Munitions Building built in 1918. It housed 9,000 War Department employees.

In Kansas City, I saw the impressive spire of the Liberty Memorial rise above the skyline while running. In 2004, Congress designated the Liberty Memorial as the National World War I Museum. The northern edge of the memorial bears a plaque stating the Liberty Memorial site was dedicated in November 1921. After the horrible sacrifice of a World War, this city decided to commemorate the war so soon afterward. Kansas City raised $2.5 million in under two weeks in 1919. That's thirty-four million in today's dollars. 200,000 people attended the dedication, including General Pershing and Marshall Foch, who were embroiled in the Great War just a few years earlier. It was a memorial to the war to end all wars. But running by so many memorials has taught me this: We haven't ended war.

When the Liberty Memorial construction was completed in 1926, President Calvin Coolidge said, "It has not been raised to commemorate war and victory but rather the results of war and victory which are embodied in peace and liberty."[1] Running up the imposing steps to the North Wall, I stare at its Great Frieze, carved by Edmund Amateis, a WWI veteran, in 1935. Coolidge's words seem innocent and hopeful. The idea that an era of peace would prevail seems almost naïve, but I think he was talking about memorializing the results of the war, not the war itself.

In Toronto, I run straight downhill toward Lake Ontario's shore and take a right for my standard five-miler. I head west on the Waterfront Trail to Coronation Park. I leave the urban canyons and go past a dog park into a stand of mature maples along the water. A massive Canadian flag with its red maple leaf flaps in the breeze. Under the flag, the Victory Peace installation marks the 50th anniversary of the end of World War II. Those six acres of maples? In Coronation Park, an oak tree was planted on May 12, 1937, commemorates the coronation of King George VI. Seven maple trees were planted around that oak that same day to represent the British Empire. More trees were planted to represent the Canadian Expeditionary Force's four divisions in the Great War. They planted almost 150 maple trees that day with each tree representing a military unit in that war. At the base of each tree, a granite stone held a brass plate that named the unit the tree represented. Few of the plaques are left, but the trees are still there.

Before researching it, I used to think it was just a nice stand of maples.

My inquiries into WWI memorials helped me learn more about other American wars. In August 2018, I took my family to the Harvard campus. We visited a building called Memorial Hall. It's an imposing structure with a layout much like a cathedral. In a portion of the building called the transept, there are massive stained-glass windows and dark wood accents. Along the walls, the names of Harvard graduates who died in the Civil War are engraved in stone panels set into the walls. One of them is my relative, Charles P. Chandler, a major in the 1st Regiment Massachusetts Volunteer Infantry. He died on June 30, 1862 at the Battle of Glendale in Virginia. His letter promoting him to lieutenant colonel arrived that day. The irony.

I visited a cemetery in Dover-Foxcroft, Maine just days later to visit an obelisk engraved with his name. They never found his body. Two memorials in two states for one man who is missing. The impulse to commemorate is strong.

The desire to honor the Civil War dead was so strong that a committee with eleven members was appointed to decide on a memorial in May 1865, just a month after Appomattox. The first stone was laid in 1870 and it was completed in 1878. Inside, we found ourselves whispering as our footsteps echoed up to the towering ceiling. A stunning massive enterprise happened so quickly after nationwide destruction.

I've started to pay attention to where to find remnants of WWI in my daily life, to know more about the powerful history of memorialization.

In September 2018, The Global War on Terrorism (GWOT) Memorial Foundation announced a project to determine the shape and location of a memorial to our ongoing wars, which brought things together in my head.

Congress authorized the GWOT memorial in August of 2017 to be on National Mall. They want to break ground in 2022 and dedicate it in 2024. The U.S. Code now says: "Commemorative works to a war or similar major military conflict may not be authorized until at least 10 years after the officially designated end of such war or conflict." Congress waived that requirement when they approved the effort to commemorate the GWOT. Watching events unfolding in Afghanistan makes me wonder when and if the GWOT will ever be over. The foundation has established six themes for the design competition, one of which is "unfinished." Effectively, there is no "officially designated end."

I've run past Pershing in DC. In Kansas City, I passed a statue of "The Hiker," which commemorates the American soldiers that fought in the Spanish-American War, the Boxer Rebellion, and the Philippine-American War. Every time I run toward the lift bridge on the Lakewalk in Duluth, I run past the Northland Vietnam Veterans Memorial and just yards later, I pass the Korean Veterans Memorial. In Canada, not all those maples were planted to commemorate World War I. Some were planted to commemorate the Fenian Raids of 1866, the 1885 Northwest Rebellion, and the Boer War. All these memorials have one thing in common. They were built after the fighting was over.

On August 4, 2017, *Military Times* wrote about the proposed GWOT Memorial:

> Supporters have argued that because the ongoing "war on terror" is open-ended, it could take decades before an official end is declared. By then, they worry, a generation of warfighters may have already died without seeing a tribute to their service.[2]

I deployed three times to Iraq and once to Afghanistan. On my third Iraq trip, I was the commander of my fighter squadron. I can only speak for myself, but the thing I worry about is people dying in war. When I was leading people in combat, it was my greatest concern. I also want us to meet our stated wartime objectives. I think all of our energy should be spent on that. Then we can worry about bringing our people home. I'm not interested in whether people pay tribute to my service. And I certainly wasn't while I was busy fighting.

That day, I continued running in Kansas City until I gravitated to the base of the 217-foot tower of the National World War I Memorial. Four Guardian Spirits are carved into the top: Honor, Courage, Patriotism, and Sacrifice. But two other confusing, striking sculptures flanking the base drew my attention. Hybrid, mythical animals with wings that encircled their faces, hiding them from the viewers. I took several pictures of those two sculptures. What were those two winged, four-legged animals with the long tails supposed to be?

As usual, I went back to my hotel room to look them up on the Internet. They're Assyrian sphinxes made of limestone and weigh 615 tons each. According to Harold Van Buren Magonigle, the architect of the Liberty Memorial:

> "Memory" located on the southwest side of the court hides its head to forget the pain and suffering of war. The figure faces east toward Flanders Field, the seat of war. On the southeast side, "Future" covers its head to attest to the skepticism of things to come and faces west where "the course of Empire takes its way."[3]

An Assyrian sphinx has the body of a bull or a lion and the wings of a bird. But here's the thing: it has a human head. The wings on the sculptures were hiding human faces. This fact, combined with the meaning behind "Memory" and "Future" resonated with me. It felt like humanity hiding its head in the sand.

I looked at my picture of the afternoon sun over "Future" as I looked toward "Memory" in the west. The wings of those two sphinxes cover human faces that won't look back and are unwilling to look forward. I think these two sculptures from the Great War already serve as a fine GWOT tribute until the current wars reach an officially designated end.

Eric Chandler is the author of *Kekekabic* (Finishing Line Press, 2022) and *Hugging This Rock: Poems of Earth & Sky, Love & War* (Middle West Press, 2017). Chandler was nominated for a Pushcart Prize in 2014 for creative nonfiction. He's a three-time winner of the Col. Darron L. Wright Award for poetry. A retired U.S. Air Force veteran of the active duty and the Minnesota Air National Guard, he flew 145 combat missions and over 3,000 hours in the F-16. Eric is a husband and father who cross-country skis as fast as he can in Duluth, Minnesota.

Notes

[1] Calvin Coolidge, "Address at the Dedication of the Liberty Memorial at Kansas City, Missouri," (The American Presidency Project, November 11, 1926), https://www.presidency.ucsb.edu/documents/address-the-dedication-the-liberty-memorial-kansas-city-missouri.

[2] Leo Shane III, "Congress Approves Plans to Build a Global War on Terror Memorial," Military Times (Military Times, August 4, 2017), https://www.militarytimes.com/veterans/2017/08/04/congress-approves-plans-to-build-a-global-war-on-terror-memorial/.

[3] J. E. McPherson and Magonigle Harold Van Buren, "Description of the Memorial" in *The Liberty Memorial in Kansas City, Missouri* (Kansas City, MO: Liberty Memorial Association, 1929), p. 28.

F. Scott Fitzgerald and WWI:
The "Crack Up" Essays

COLIN D. HALLORAN, veteran

Paul Fussell, in his seminal *The Great War and Modern Memory*, posits that "logically, one supposes, there's no reason why a language devised by man should be inadequate to describe any of man's works. The difficulty was in admitting that the war had been made by man and was being continued ad infinitum by them."[1]

While there is much debate and discussion over the official definition and dates of Modernism, we cannot overlook WWI and the ways it changed literary language. Broadly, the Modernist movement sought to move away from traditionalism and towards originality, particularly focusing on a "nonlogical, non-objective, and essentially causeless mental universe."[2]

Because the war itself was nonlogical. Even the innovative language and stylizations that propelled Modernist writings prior to the war were suddenly inadequate after the horrors the world now knew humankind was capable of inflicting on itself.

Yet much of the poetry to come out of WWI was still focused on the collective "we" and the broader identifiers like "English," "American," "French," "German," "Homefront," and "Trenches." Non-fiction remained external—largely historical and fact-based. Fiction writers, on the other hand, delved into the internal workings of the individual mind. For example, Freud's work with WWI veterans and dreams helped fuel the movement's interest in the human subconscious and psyche, leading writers to approach their realities and experiences through metaphor, mythology, internal monologues, and even dream sequences, as in Wilfred Owen's "Strange Meeting."

In addition, the Great War stripped young authors—many of whom would shape the Modernist movement of interwar literature—of their idealism. Included in this group was titan of the twenties, F. Scott Fitzgerald, author of *The Great Gatsby*. Fitzgerald, in spite of dropping out of Princeton to join the Army as a second lieutenant, never shipped out, a fact that he would later lament.

Unlike many Modernist authors of the time who were pushing the boundaries of fiction with experimental forms and techniques, Fitzgerald and his contemporary, Ernest Hemingway, kept their writing in the realm of realism that was so popular in the 19th century. However, their stylization, characters' attitudes, and choice of themes place them firmly within the Modernist oeuvre. For example, while there is no question that Hemingway's fiction is highly autobiographical, he was able to distance himself from his own experiences by assigning them to his various characters, such as WWI ambulance driver Frederic Henry in *A Farewell to Arms*. This technique surely contributed to his success, as many of his readers recognized their own thoughts and experiences in the musings of Hemingway's fictional narrators. Hemingway and his work embodied the values of stoicism and ambivalence that were to be expected from a world emerging from the devastation of war. Boys had become men and men had died doing their duty, serving their homelands, protecting what was right and good, as extolled in so many poems and media of the time. Detachment was viewed as strength, and strength was now expected.

Which is also why some lesser-known works by Fitzgerald are so important.

I am referring especially to the so-called "Crack-Up" personal essays published in *Esquire* in 1936. The first essay sets the fragmented, dismal tone of the collection; it begins, "Of course, all life is a process of breaking down . . . "[3]

Many of Fitzgerald's contemporaries and friends recoiled at these autobiographical, emotional essays that chronicled his own personal post-war crisis. In fact, as if embarrassed for his friend, novelist John Dos Passos, wrote to Fitzgerald:

> . . . most of the time the course of world events seems so frightful that I feel absolutely paralysed [sic] . . . We're living in one of the damndest tragic moments in history—if you want to go to pieces I think it's absolutely O.K. but I think you ought to write a first rate novel about it . . . instead of spilling it in little pieces.[4]

Dos Passos' statement reflected the ways American society shunned the display of male emotion. In the post-war years, male emotion was acceptable for public consumption only if it were fictionalized, not as an autobiographical confession.

Fitzgerald explored this very dilemma in the essay, "Sleeping and Waking," published just more than a year before *The Crack-Up* essays appeared in print. Written in response to Hemingway's short story, "Now I Lay Me," which discusses the agony of insomnia, Fitzgerald observes "there was nothing further to be said about insomnia...it appears that every man's insomnia is as different from his neighbour's as are their daytime hopes and aspirations."[5] Fitzgerald acknowledges here that "Now I Lay Me" must be a fictional representation of Hemingway's own struggles with insomnia and emotional suffering. But, in recognizing that insomnia is deeply personal, he also seems to say that there is a need for further, public discussion on this kind of suffering, perhaps directly, rather than filtered through fiction.

Fussell points out that in real correspondence, WWI soldiers in the trenches would make use of the passive voice to create a sort of narrative distance. For example, Fussell says that a soldier might write in passive voice "A very odd sight was seen here," instead of using the active voice with "I saw [...]" to "avoid designating themselves as agents of nasty or shameful acts."[6]

The deployment of fiction served the same purpose: distancing the author from real emotion. Fictionalizing acts and thoughts and psychological struggles allowed the author to maintain a sense of credibility in the public's eyes. It allowed them to maintain the masculine façade of the early twentieth century. It allowed them to maintain a perspective gained and solidified during the war years, no matter how long after it they were writing. As Fitzgerald himself observes in his essay "Ring," "A writer can spin on about his adventures after thirty, after forty, after fifty, but the criteria by which these adventures are weighted and valued are irrevocably settled at the age of twenty-five"[7]

This reminds me of Tim O'Brien's "How to Tell a True War Story," and John McCrae's "In Flanders Fields." As a war veteran, I know firsthand that the rest of soldiers' lives are shaped by what we were taught on the way to war, in war, and as we made that terrible transition back into a civilian world that lauded those "nasty or shameful acts" as heroic. These experiences shape the way we view all those that follow, whether we want to acknowledge them or not.

Fitzgerald explores these ideas later in his essay "Sleeping and Waking" as he recounts "the war dream." To get himself to sleep, he replays a fabricated scenario in which the Japanese have invaded the U.S. and made it as far as Minnesota, Fitzgerald's home state. In reality, he was never deployed, but, in the dream, he is an assistant to a general and called "Captain Fitzgerald." He is "the character who bears [his] name has become blurred," but, in the dream, it is also vital to the survival of his men and his country. As he recounts the heroics of defending the place he knows so well, Captain Fitzgerald's tale falls apart. He writes, "waste and horror—what I might have been and done that is lost, spent gone, dissipated, unrecapturable. I could have acted thus, refrained from this, been bold where I was timid, cautious where I was rash. I need not have hurt her like that. Nor said this to him. Nor broken myself trying to break what was unbreakable."[8] We see, perhaps for the first time—in fiction or personal essay—the guilt Fitzgerald bears because he was never deployed during the Great War.

This is a feeling that many veterans I know share. Not having deployed, or having deployed but not left Kuwait or Bagram or the FOB. Or having deployed but only once or twice when that guy there had deployed seven times. Or, in my case, having been medevacked out and living with the guilt of leaving my men behind, even though I had no say in the matter. And just like Fitzgerald, this particular guilt of war is so often the launching point, the pry bar that opens the vast fissure of all other guilts in my life.

But I'm getting ahead of myself, and Hemingway would not approve. At least Fitzgerald might.

In the opening lines of "The Crack-Up," Fitzgerald states, "Of course all life is a process of breaking down, but the blows that do the dramatic side of the work—the big sudden blows that come, or seem to come, from the outside—the ones you remember and blame things on and, in moments of weakness, tell your friends about, don't show their effect all at once."[9] This sentence, packed with dense ideas, changed the way Fitzgerald's contemporaries viewed him and perhaps the way he viewed himself; he was no longer the debonair socialite with an acute talent for observing human behavior in social interactions but a broken, troubled man.

First comes the idea of the internal versus the external with "the big sudden blows . . . that seem to come from the outside."[10] The majority of WWI and interwar literature focused on factors external to the human psyche: facts, figures, patriotism, illustrations of events and scenes, while Modernists focused on the internal process that happens long after such events take place: "the ones" that "don't show their effect all at once."[11] Second, it hints at the idea of life as a continuous process of breaking down, which, if applied globally, speaks to war as a natural part of humanity, which goes against Fussell's claims of the inadequacy of language to describe it. It also speaks to the idea of key moments in life lingering, not showing their full impact all at once, which contextualizes Fitzgerald's own view that a writer's viewpoints are solidified by the age of twenty-five. But perhaps most importantly, it points out that the act of acknowledging that you are breaking down, even to one's closest friends and confidantes, is the result of weakness, the antithesis of early twentieth-century masculine values, and one of the hurdles contemporary military writers like me still face.

Part of this could be attributed to the trend we see in the poetry of WWI of the collective "we," the soldier as a part of an idea, something bigger than himself. This foundational concept is still an essential part of military tradition: the stripping of the individual, the reshaping of a recruit into part of a unit, a piece of a greater whole. Uniform. Maybe this is why *Full Metal Jacket* remains one of the most popular war films among veterans; it focuses on how a soldier is made rather than on what they are intended to do.

In the second "Crack-Up" essay, "Pasting it Together," Fitzgerald parallels this concept by considering those who influenced him most. His ultimate conclusion? "So there was not an 'I' anymore—not a basis on which I could organize my self-respect—save my limitless capacity for toil that it seemed I possessed no more. It was strange to have no self." Such realizations, he acknowledges, "always confused me and made me want to go out and get drunk."[12]

These sentiments, the breaking down of individual identity, and the realization there is a lack of a true self are elements that are at the forefront of contemporary military literature. How does one balance their own personal feelings and beliefs with the mission of the machine in which they are a cog? How do we reconcile the idea that the many influences in our lives may, at times, be in conflict? How do we, as veterans especially, find an individual identity when we leave the collective "we" of the military?

These are all questions that Fitzgerald raises and that come up in post-9/11 war literature. Another commonality? The solution: it "made me want to go out and get drunk" because it's damn confusing.[13] These are themes that run throughout much of the personal creative-nonfiction coming out of these current wars. How do we reconcile our roles as warfighters once we're back in the realm of civilians?

The answer to this question is something Fitzgerald gets closer to in the third and last of his "Crack-Up" essays:

> My self-immolation is something sodden-dark. It was very distinctly not Modern—yet I saw it in others, saw it in a dozen men of honor and industry since the war . . . I had stood by while one famous contemporary of mine played with the idea of the Big Out for half a year; I had watched when another, equally eminent, spent months in an asylum unable to endure any contact with his fellow men. And of those who had given up and passed on I could list a score.[14]

At its core, "The Crack-Up" is an account of how "an exceptionally optimistic young man experienced a crack-up of all values, a crack-up that he scarcely knew of until long after it occurred."[15]

Again, this is a sentiment expressed throughout the literature of the Global War on Terror. On September 11, 2001, the world changed. For the first time since WWII young people, the government, and the citizenry, had a clear "them" to our "us." Though this new war would not have the clarity of blue uniforms against grey uniforms, of the khaki and green of the allies against the grey of the Germans that WWI poets used to subvert Romantic tropes, there at least was a clearly defined good versus evil. Or at least that's how it seemed at the start, at least that was the sentiment that caused so many young Americans, including myself, to enlist.

Fitzgerald's final essays in the 1930s finished with a great deal of cynicism. Gone was the hopeful—if arrogant—character readers met in *This Side of Paradise,* Amory Blaine, a reflection of Fitzgerald in fiction that would make even Hemingway proud. In his place was a man, a writer, a patient, who saw the value in no longer fictionalizing his truth. In his place a man who, rather than seeking the false glories of war, declares:

> Let the soldiers be killed and enter immediately into the Valhalla of their profession. That is their contract with the gods. A writer need have no such ideals . . . the old dream of being an entire man . . . has been relegated to the junk heap of the shoulder pads worn for one day on the Princeton freshman football field and the overseas cap never worn overseas.[16]

Fitzgerald never became the Princeton football star he dreamed of. He never became the leader of military men who he tried to use to find rest years after the war had ended.

And yet, Fitzgerald leaves his readers with an odd parallel: "My own happiness in the past often approached such an ecstasy that I could not share it even with the person dearest to me but had to walk it away in quiet streets and lanes with only fragments of it to distil into little lines in books."[17] Once again, we see the repression of emotion, though this time it's happiness. What we're left with, is the conclusion that the men of the interwar years we're supposed to be stoic. That trials and triumphs alike were to be suppressed, even to those they were closest to. And yet in the last decade of his life, Fitzgerald—who had never gone to war, but who wished to and who moved in circles of those who had—wrote about what the previous decade had caused him to feel, to do. He gave voice—real voice, not fictionalized—to what so many of his contemporaries were feeling. What so many of his contemporaries were not allowed by society to write as true experience. Though he expressed what so many had translated to fiction—what so many people saw, but ignored in society—he was shunned, berated, dismissed, even by those closest to him.

And yet, without his writing, without the death of his dreams, contemporary war writing would not be where it is today. Confessional poetry would emerge after WWII. The Vietnam War would see poetic narrative voice shift from the collective, nationalistic "we" to the confessional, shamed "I." But it would not be until post-9/11 war literature that Fitzgerald's influence would truly be felt.

This man, who would receive so much flak from me and my combat arms brethren today for being a POG, paved the way for us to write honestly about our experiences in poetry and non-fiction. His bravery in alienating himself from his peers preempted so many war writers of the last fifteen years who have been questioned, criticized, and ridiculed by their peers. Much of this negative response has been directed toward female veterans, but the truth is that male veterans experience this underlying conflict just as much, if not more.

The difference is men are perhaps more hesitant to reveal the true impact of their experiences. It is my hope that they will turn to Fitzgerald as an example. That they will do what I and so many of my peers have done and reveal the true impact of war—whether deployed or not. It is my hope that civilians and veterans alike will share their true stories, not bury the emotional impact of war in fiction.

And it is my hope that our peers, those whose voices can't reach the public eye, are able to learn from the courage of F. Scott Fitzgerald and share their stories.

And it is my hope that those who are not able to share, willing to share, or allowed to share because they are still in the military, will recognize their own stories in ours.

But most importantly, it is my hope that in these wars, a century after the cessation of hostilities in WWI, we can all—military or civilian—recognize the value in sharing not just the external truths of war, which is something modern media allows us to experience daily, but also in sharing the true psychological impact of war. To recognize those sleepless nights, those many looking for the "Big Out," that drive to drink, and that loss of self. Let us, as writers, follow the path Fitzgerald laid for us, and let us, as readers, not respond as Hemingway and Dos Passos, but as citizens who recognize that it is we who have sent these young men and women to their own particular "Crack-Up."

Colin D. Halloran is an Army veteran of Afghanistan and author of the poetry collections *Shortly Thereafter*, a verse memoir of his service in Afghanistan, *Icarian Flux*, which uses persona to delve into the complexities of life with PTSD, and *American Etiquette*, which explores the consequences of American violence at home and abroad. He has had essays published in collections such as *Retire the Colors: Veterans & Civilians on Iraq & Afghanistan* and *Why We Write: Craft Essays on Writing War*. When not writing, Halloran gives talks on trauma and leads workshops that seek to heal trauma through the arts.

Notes

[1] Paul Fussell, *The Great War and Modern Memory* (London: Oxford University Press, 2000), 170.

[2] William R. Everdell, *The First Moderns: Profiles in the Origins of Twentieth-Century Thought* (Chicago: University of Chicago Press, 1998), 11.

[3] F. ScottFitzgerald, "The Crack-Up," in *The Crack Up*, ed. Edmund Wilson (New York: New Directions, 2009), 69.

[4] John Dos Passos, "A Letter from John Dos Passos: October 1936," in *The Crack-Up*, ed. Edmund Wilson (New York: New Directions, 2009), 311.

[5] F. Scott Fitzgerald, "Sleeping and Waking," in *The Crack-Up*, ed. Edmund Wilson (New York: New Directions, 2009), 67.

[6] Paul Fussell, *The Great War and Modern Memory*, 177.

[7] F. Scott Fitzgerald, "The Ring," in *The Crack-Up*, ed. Edmund Wilson (New York: New Directions, 2009), 36.

[8] Fitzgerald, "Sleeping and Waking," 67.

[9] Fitzgerald, "The Crack-Up," 69.

[10] Fitzgerald, 69.

[11] Fitzgerald, 69.

[12] F. Scott Fitzgerald, "Pasting it Together," in *The Crack-Up*, ed. Edmund Wilson (New York: New Directions, 2009), 79.

[13] Fitzgerald, "Pasting it Together," 79.

[14] F. Scott Fitzgerald, "Handle with Care," in *The Crack-Up*, ed. Edmund Wilson (New York: New Directions, 2009), 81.

[15] Fitzgerald, "Handle with Care," 80.

[16] Fitzgerald, 83-84.

[17] Fitzgerald, 84.

History Between Humor and Tragedy:
Musings on Robert Graves' Memoir,
Goodbye to All That*

DAVID JAMES, veteran

When thinking about WWI, I am reminded of a quote by Jorge Luis Borges about the 1982 Falklands War, "It is a fight between two bald men over a comb."[1] In a similar way, we could say that WWI was a fight between a bunch of spoiled children over who got to use the playroom. Though they all had their toys, sharing and cooperation were unlearned traits. It is important to remember this tragedy, though sometimes the easiest way to deal with tragedy, if not by outrage, stoicism, or escapism, involves a disarming sense of humor and irreverence. Robert Graves' *Goodbye to All That*, his memoirs of early life in England, his participation in the trenches of WWI, and his post-war experiences encompass various coping mechanisms, including humor.

A prolific poet and author, Graves is most famous for his fictional rendering of the Julio-Claudian dynasty in I, *Claudius and Claudius the God*. Born in 1895, Graves was nineteen when the war began—a typical age for new officer and soldier recruits. His mother was German, and his middle name was von Ranke, which was no small problem considering the bullying, nationalistic, anti-German WWI-era hysteria that caused suspicion from schoolmates and later from fellow soldiers despite his proven competence in battle. As a pacifist married to a German who was under de facto house arrest for the entire war, writer D.H. Lawrence experienced this on a smaller scale.

Goodbye to All That, published in 1929, eleven years after the Armistice, was Graves' second work of nonfiction after a biography of his friend T.E. Lawrence, *Lawrence and the Arabs*. By this time, Graves had also published many poetry collections. The publication of his memoirs came when the young author had just recovered from years of emotional trauma that today we would call PTSD ("shell shock"), and the title references what he calls his "bitter leave-taking of England"

[2]— its war, politics, society, education, and even many of his own family and friends. He writes of his post-war experience:

> Very thin, very nervous, and with about four years' loss of sleep to make up, I was waiting until I got well enough to go to Oxford on the Government educational grant. I knew that it would be years before I could face anything but a quiet country life. My disabilities were many: I could not use a telephone, I felt sick every time I travelled by train, and to see more than two new people in a single day prevented me from sleeping. I felt ashamed of myself as a drag on Nancy but had sworn on the very day of my demobilization never to be under anyone's orders for the rest of my life. Somehow, I must live by writing.[3]

After publication of *Goodbye to All That*, Graves escaped to the Spanish island of Majorca where he remained for the rest of his life, except for a long stay in America to flee the Spanish Civil War.

Goodbye to All That is important for its ability to capture, from the point of view of an individual soldier rather than from a comprehensive historian, the passing of one epoch to another. From WWI, which has been called the "long 19th century" or the "belle epoque" came the "modern age."[4] These are mere historical categories, but they serve to reflect the turbulence brought on by changes in the old-world system dating from before the French Revolution, to a new world where possibilities for progress and destruction both expanded exponentially.

Graves serves as a paradigm of a young, well-educated person from England's middle class who, like many throughout the West, felt reshaped after WWI. Their thoughts shifted towards more radical ideas including socialism, atheism, feminism, and pacifism precisely because of their experiences in the trenches. They came out disillusioned with a new view of a society that they discovered to be neither as civilized nor as progressive as they had thought. Thomas Mann in *The Magic Mountain*, for example, captures this shift from the German perspective.

Graves opens his memoirs with a conventional account of his family history and early years. He mentions witnessing Queen Victoria's 1897 Jubilee and he spends some time in these early chapters detailing his visits to his aristocratic German relatives in their Bavarian castles, relatives who would later become his enemy.

He attended many public schools in England, which Americans call private or prep schools, spending the longest times at Charterhouse. Several anecdotes denounce the severity and hypocrisy of this education system. These outdated but still powerful Victorian standards of morality accomplished little more than to stifle emotional development and even foster what would have been considered at the time immoral. He talks about the prevalence of homosexuality in these types of all-boy boarding schools, going so far as to detail his own platonic infatuation with a younger schoolmate. He also dwells on his friendship with George Mallory, the famous alpinist who was an older mentor at Charterhouse and later best man at Graves' wedding. He mentions Mallory, who died on Mount Everest in 1924 after possibly becoming the first person to reach the summit, as one of the only people who treated students like humans, which puzzled everyone. We also learn at this time that Graves took up boxing as much to defend against bullies as to keep fit, and this would later prove useful in demonstrating his manliness in front of soldiers and superiors alike.

The heart of the book comes in the middle chapters detailing his time spent on the Western Front. At the outbreak of war, he deferred his matriculation to Oxford University to join the army. He was commissioned as a lieutenant in the Welsh Regiment since his family home was in Harlech in northwest Wales. Like other young men, he was eager to join in the fighting before the war ended.

How many times has it been said that a war will be over by Christmas?

While the war obviously did not end by December 25, 1914, Graves witnessed the famous Christmas Day truce soon after joining his regiment on the Western Front. He refers to it as the Christmas 1914 fraternization, and his regiment was among the first to participate. This event, rare in the annals of war, saw the belligerents, German, French, and British, come out of their trenches and join in an unarmed singing of carols and exchange of greetings and gifts. This short-lived sense of shared humanity and brotherhood shows soldiers who shed their martial spirit and attempt to take back control of some part of their lives outside of battle.

I spent two Christmases in Afghanistan and understand well the sentiment that comes at times like holidays when all we desire is a temporary break from the stress and trauma of war. Even in 1914, the truce was obviously resented by the generals and politicians, who ensured there would not be a repeat of such non-warlike sentiment the next Easter or following Christmases. They also convinced the press in the involved countries to keep quiet, where no mention was made for at least a week after the 1914 truce. No one knew right away that hundreds of thousands laid down their arms to hobnob with the enemy. And when it was finally covered, the press distorted and minimized the truce to make it seem more freakish and less peaceful. The Christmas Day truce lives on in popular memory and culture. In 2014, the British supermarket Sainsbury's went so far as to make a television commercial reenactment of it in which a German and British soldier swap chocolate and biscuits.

One of the central events in *Goodbye to All That* is the Battle of Loos, a British and French attack on German lines in September 1915 when only a few kilometers of ground changed hands and almost 100,000 men died. It marked the first use of poison gas by the British, and also chronicled the battle in which Rudyard Kipling's son went permanently missing in action, prompting the author of *The Jungle Book* to write the sad poem "My Boy Jack."

Graves describes how the gas was euphemistically referred to "the accessory,"[5] and how everyone was highly skeptical of its efficacy because the supervisors were university chemistry professors brought in to administer it. Sure enough, "the accessory" was deployed with a headwind coming into the Allied lines, causing the gas to harm the British more than the Germans for whom it was intended.

The Battle of Loos was a disaster for him. Graves writes that, much later in the war when he had been sent home to recover from his wounds, he was asked to give a speech to 3,000 incoming Canadian soldiers. He writes:

> They were Canadians, so instead of giving my usual semi-facetious lecture on 'How to be Happy, Though in the Trenches', I paid them the compliment of telling the real story of Loos, and what a balls-up it had been, and why – more or less as it has been given here. This was the only audience I have ever held for an hour with real attention. I expected Major Currie to be furious, because the principal object of the Bull Ring was to inculcate the offensive spirit; but he took it well and put several other concert-hall lectures on me after this.[6]

A key feature of *Goodbye to All That* is the farcical and probably invented dialogue, which reads like short theatrical set-pieces. It seems like almost every occasion of reported speech involves a back-and-forth rhythmic dialogue that ends in someone laying a punchline. Along with the stock characters, we grasp the fictionalized nature of Graves' memoirs.

Siegfried Sassoon is one of the most important characters in Graves' book. He is a fellow war poet who joined Graves' Royal Welch Fusiliers regiment in 1916. They struck up an immediate friendship. Sassoon published his own three-part fictionalized autobiography in the 1930's, *Memoirs of an Infantry Officer.* Like Graves, Sassoon had not published any poetry when they first met, and Graves' realistic style influenced his friend. Graves describes Sassoon as one of the most courageous men he had ever seen or heard about in his time in the trenches. He tells one story about how Sassoon single-handedly attacked and took control of a German observation trench, then enraging his superiors by not telling anyone about it. He was found two hours later sitting in the German trench reading a book of poetry.

Sassoon, like Graves, later suffered a type of nervous breakdown and wrote his famous 1917 "Soldier's Declaration," denouncing the war and the government's incompetent prosecution of it. He was encouraged by anti-war activists like Bertrand Russell and Ottoline Morrell. Sassoon threw his Military Cross for bravery into a river, though he escaped a court-martial, with Graves' help, and was sent to a hospital to recover from shell shock. There he met Wilfred Owen, another war poet who became influenced and encouraged by Sassoon, and who died on the Western Front one week before the Armistice.

Sassoon and Owen were both gay like Austrian philosopher Wittgenstein who, like Sassoon, volunteered for service at the outbreak of war and demonstrated such incredible repeated bravery in battle on the Russian Front, he was thought to be suicidal. Such examples show why it makes no sense that gay soldiers in the American military have until recently been considered unfit for service.

One of the most tragic, and understated, events of the book occurs when three officers of Graves' battalion, along with three of his closest friends, were all killed in the same day by shelling and sniper fire. David Thomas, the third member of the trio of poet friends in the battalion, was among the dead. Graves writes:

I felt David's death worse than any other since I had been in France, but it did not anger me as it did Siegfried. He was acting transport-officer and every evening now, when he came up with the rations, went out on patrol looking for Germans to kill. I just felt empty and lost.[7]

Soon after, he says:

My breaking-point was near now, unless something happened to stave it off. Not that I felt frightened. I had never yet lost my head and turned tail through fright, and knew that I never would. Nor would the breakdown come as insanity; I did not have it in me. It would be a general nervous collapse, with tears and twitchings and dirtied trousers; I had seen cases like that.[8]

Graves finished his time in the trenches during the 1916 Battle of the Somme, when he was injured so gravely that he was first reported as dead. He spent the rest of the war convalescing in hospitals, helping train new volunteers to his unit, and was even posted to Ireland where the English garrison was trying to stop the burgeoning Irish uprising. The rest of the book talks about his marriage to a feminist activist, their move to the country near Oxford, setting up house and opening a general store. He admits:

The moral problems of trade interested me. Nancy and I both found it very difficult at this time of fluctuating prices to be really honest; we could not resist the temptation of under-charging the poor villagers of Wootton, who were frequent customers, and recovering our money from the richer residents. Playing at Robin Hood came easily to me. Nobody ever detected the fraud.[9]

He had four children in eight years and, he describes what amazes me most about his autobiography—sometimes he would only scrape out half an hour or so of writing a day in between his fatherly and household caretaking duties!

In this later part he also deals at length with his friend and biographical subject, T.E. Lawrence. He relates his post-war experience to Lawrence's:

I knew nothing definite of Lawrence's wartime activities, though my brother Philip had been with him in the Intelligence Department at Cairo in 1915, making out the Turkish Order of Battle. I did not question him about the Revolt, partly because he seemed to dislike the subject—Lowell Thomas was now lecturing in the United States on 'Lawrence of Arabia'—and partly because of a convention between him and me that the war should not be mentioned: we were both suffering from its effects and enjoying Oxford as a too-good-to-be-true relaxation. Thus, though the long, closely-written foolscap sheets of The Seven Pillars were always stacked in a neat pile on his living-room table, I restrained my curiosity. He occasionally spoke of his archaeological work in Mesopotamia before the war; but poetry, especially modern poetry, was what we discussed most.[10]

The book ends in 1929, though shortly after he divorced his first wife, Nancy, and got remarried and had four more children with his poetic muse, Laura Riding, with whom he established a publishing company at their base on Majorca. He was runner-up for the Nobel Prize in Literature, which Steinbeck won that year, and he died at the age of 90 with 140 published works.

Graves' memoirs are filled with stories of understated humor and pathos. In one case, he describes the last time he attended church, during his Easter 1916 visit home. He tells the story of having to push his mother uphill in a heavy bath chair, since the only available wheelchair in town was taken by "Countess of-I-forget-what," and then sits through a three-hour service despite being ill. During the sermon the "strapping" young curate, one of four men present—compared with seventy-five women—was "bellowing about the Glurious Performances of our Sums and Brethren in Frurnce today. I decided to ask him afterwards why, if he felt like that, he wasn't himself either in Frurnce or in khurki."[11] His father then took him to meet War Secretary (and future Prime Minister) David Lloyd-George, who Graves says "sucked power from his listeners and spurted it back at them. Afterwards, my father introduced me to Lloyd George, and when I looked closely at his eyes, they seemed like those of a sleep-walker."

Graves' book angered so many people that even his father, one of the offended, felt it necessary to write his own memoirs as a rebuttal to his son's entitled *To Return to All That*.

It could be said that Graves was playing with fire. Developing a sense of humor toward a destructive war declared by elites but suffered by the common man, is, I think, not only in bad taste but can do more harm than good because it normalizes the illegality and immorality of the war. Thus, I agree with this quote by Bertrand Russell, a pacifist who spent the last year of World War One in prison for speaking against involuntary military service for conscientious objectors:

> Alas, I am that extremely rare being, a man without a sense of humour. I had not suspected this painful fact until the middle of the Great War, when the British War Office sent for me and officially informed me of it. I gathered

that if I had had my proper share of a sense of the ludicrous, I should have been highly diverted at the thought of several thousand young men a day being blown into tiny little bits, which, I confess to my shame, never once caused me to smile. I am reminded of a Chinese emperor, who long ago constructed a lake made entirely of wine, and then drove his peasants into it only to amuse his wife with the struggles of their drunken drownings. Now he had a sense of humor.[12]

Yet, at the same time, the dark or a cynical sense of humor displayed by veterans against their war may also be a way to ease the personal trauma and represent, in a fictionalized way, the collective tragedy in which they played a part. As such, I look up to Graves and his successors such as Joseph Heller and Kurt Vonnegut, who have highly influenced the field of war literature.

As for the causes of destructive and self-destructive wars like WWI, I will leave it again with the wise and quotable Bertrand Russell, and his book *Education and the Social Order*. Russell writes of the innate violent sense of retributive justice that is easily awakened in humans: "I found one day in school a boy of medium size ill-treating a smaller boy. I expostulated, but he replied: 'The bigs hit me, so I hit the babies; that's fair.' In these words, he epitomised the history of the human race."[13]

Perhaps one of the things that makes us human is the ability to laugh in the face of the tragically absurd and continue living on in spite of it. Graves accomplishes this in his book, *Goodbye to All*. It is a classic not only in the genre of war literature but for all modern literature that deals with tragedy.

* The *Wrath-Bearing Tree* has given MilSpeak permission to republish this essay, which first appeared under the title of "Goodbye to Christmas Truces" on December 24, 2014.

David James served in the U.S. Army in Afghanistan from 2005-06 and 2007-08. He now teaches history in Northern Italy, where he lives with his wife and twin daughters. He agrees with Borges that "reading is an activity subsequent to writing more resigned, more civil, more intellectual."

Notes

[1] Luis Jorge Borges quoted in *Time* (Vol. 121, No. 7), February 14, 1983.

2 Robert Graves. *Good-Bye to All That: With a Prologue and an Epilogue* (New York: Doubleday, 1989), 7.

3 Graves, *Goodbye*, 317-318.

4 Eric Hobsbawn. *The Age of Revolution: Europe, 1789-1848* (New York: World Publishing, 1962).

5 Graves, *Goodbye*, 163.

6 Graves, 201.

7 Graves, 217-218.

8 Graves, 218.

9 Graves, 340.

10 Graves, 329.

11 Graves, 221.

12 Bertrand Russell. *Autobiography* (New York: Routledge, 2009).

13 Bertrand Russell. *Education and the Social Order* (Crows Nest, Australia: George Allen & Unwin, 1932), 18.

Of the Dreadnoughts

JEFFERY HESS, veteran

Florida and the Navy are as much a part of me as blood and bone. I've lived in the state all but the first four years of my life. But the only thing I find myself writing about more than Florida is the Navy. Having served abbard two ships, I'm endlessly fascinated by shipboard life—especially in times of war. I've written about Navy ships and submarines and populated them with people real and imagined. A few years ago, I was putting together a collection of Navy stories set during the Cold War. I had a good number of stories, but I always looked for the next idea, the next ship. The stories I had were heavy on the later portion of the era, so I began looking back.

I did some digging, took copious notes, and accidentally rediscovered the Dreadnoughts.

I always found it odd that my buddy played what's commonly referred to as a dreadnought guitar. And a high school in the next town calls their team the Dreadnaughts. (I still don't understand the alternate spelling.) I was familiar with the Navy's use of the word and long a fan of battleships. Though I never got to serve aboard one, I was at sea aboard a guided missile cruiser in the Caribbean with the USS *Iowa* during the turret explosion in 1989.

Still, I had no business researching WWI during my Cold War project, but the word drew me in. I couldn't resist looking into these ships.

I'd gone down the rabbit hole several times on a number of topics over the years, but for days I was wrapped up in the world of these heavily armed and mighty ships. The design of which originated in England. They were so impressive it's no wonder the U.S. Navy quickly emulated and improved the designs.

I took pleasure in seeing there was not only a dreadnought battleship named *Florida,* but also to know she was the first in her class.

Again, I was far afield from the Cold War, but right about that time, I also discovered a bunch of Navy recruiting posters from 1917 in the Library of Congress online archive. Two caught my eye.

In "America Calls - Enlist in the Navy" by J.C. Leyendecker, the Statue of Liberty shakes hands with a sailor in dress whites holding a rifle near the bayonet.

In "Join the Navy, the Service for Sighting Men" by Richard Babcock, a sailor in his winter blue uniform rodeo rides a torpedo through the water.

Being a writer means I always look for a story, and so I began wondering if there was a kid, let's call him Wyatt, from a farm of some type in a landlocked state. Now, what if Wyatt saw one or both of those posters at the post office or maybe in his principal's office at school? Maybe he saw himself as the guy in the poster with Lady Liberty's arm around him. I think they looked a lot alike.

And what if Wyatt figured he'd get off the farm and go serve his country? Given the choice of being a doughboy or a sailor those recruiting posters were all he needed to know.

The Statue of Liberty played to Wyatt's sense of patriotism. The torpedo riding poster catered to his wild side that likely had gotten him paddled a few times at school and then at home. Maybe Wyatt lied about his age and enlisted in the Navy? What if he got away with it? Maybe he would've gotten lucky enough to receive orders to one of the dreadnoughts, say, the USS *Florida*.

Commissioned in 1911, USS *Florida* (BB-30) was almost brand new when Wyatt got aboard. He must've been in awe.

She ran five hundred twenty-one and a half feet, just a little shorter than either of the ships I served aboard—eight feet and forty-six feet, respectively. Which means even someone Wyatt's below-average height had few places to stretch out aboard ship.

Aboard ship, new recruits were lined up to have their musculature assessed. Wyatt was muscular but short. He was deemed worthy to shovel coal in a boiler room twelve hours a day whether he was better suited for something else or not. And there wasn't just one hungry beast, but twelve boilers, all burning hot and demanding to be fed.

The retina-burning glare through the open furnace door got Wyatt at least once every day.

It was too hot for shirts down there. His skin was coated in coal dust, rained black sweat. Some days, at least one of the shovelers would pass out and have to go topside for a while.

Temperatures could exceed one hundred thirty degrees. These coal shoveling sailors didn't know anything about hydration and instead worked through cramps of all degrees and tried desperately to not become one of the ones who passed out. And we all know the drawbacks of inhaling coal dust over the long term.

If he had been a little taller, Wyatt might've been selected to hoist ammo shells into one of the twelve-inch guns. He could've been a loader or rammer who hoisted projectiles. Both were demanding work for even the strongest young men, but he would've enjoyed it more to contribute to explosions than tending fires no one ever saw. Yet, in my mind, a salty old chief explained the importance of his role, and Wyatt bought into the chief's words and did his job flawlessly every day all day until he could go get some rest.

Hammocks that were stowed between reveille and taps filled the berthing spaces of these ships, leaving open deck space to serve as the mess deck for the hungry crew. They ate their meals in the same space from tables that swung down from bulkheads amongst the aromas of coal residue, dirty socks, and three hundred armpits at a time.

In between meals during any length of downtime, these open berthing spaces also served as gambling halls and makeshift boxing rings for sanctioned or unsanctioned boxing matches, known as "smokers," which were held to help the men blow off steam and be entertained, which was essential for camaraderie and morale during the long and monotonous workdays at sea.

The best I figure it, Wyatt held a record of 7-1 and stayed friends with all the men he fought.

Jeffery Hess is the author of the novels *Roughhouse, Tushhog, Beachhead,* and *No Salvation*, and the short-story collection *Cold War Canoe Club* as well as the editor of the award-winning *Home of the Brave* anthologies. Prior to earning an MFA in creative writing from Queens University of Charlotte and a bachelor's degree in English from the University of South Florida, he served aboard the Navy's oldest and newest ships. He lives and writes in Florida, where he leads the DD-214 Writers' Workshop for military veterans. More at jefferyhess.com.

A Movie That Made Me:
A Farewell to Arms

JENNY PACANOWSKI, veteran

Often people have asked me at my poetry readings if I started writing before my 2004 deployment to Iraq or before my discovery that veteran writing workshops and poetry were a part of me that I couldn't live without. I've said that I didn't write before the war except in private journals. However, when recently watching old movies on my parent's couch, I realized I was probably and most nearly born a poet.

One of the movies that's inspired me was based on the novel by Ernest Hemingway, *A Farewell to Arms*. A volunteer ambulance driver, Frederic Henry, and a nurse, Catherine Barkley, meet and fall in love in Italy during the Battle of Caporetto in WWI.

While the time and circumstances were different, their situation 100 years ago resonates with my experience as a medic in Iraq. The 557th medical evacuation unit I was assigned to was split up and sent throughout the country to different bases and forward operating bases to support convoys, hospitals, and small medical stations, which probably is why I related more to Frederic as an itinerant ambulance driver than to Catherine as a hospital nurse. I did medical support for convoys and watching Frederic drive towards the sounds of bombs and guns was a familiar action. I understood deeply why he had previously been in our terms today, "a player," because when you love someone in a combat zone, the choice to go towards danger instead of away from it becomes difficult. The desire to return to base becomes your driving force.

Like Frederic, my lover was also in a base hospital—in charge of the corpsmen at a Navy hospital—to which my platoon was attached at Al Asad Airbase. Going back to base at night and sneaking into his room counts as one of the only times I felt content in Iraq. Lying in the arms of someone you care about is truly consoling after facing death. It makes war less lonely. I saw that in the intense encounter between Catherine and Frederic.

Catherine's rage stands out to me when she thinks that Frederic was insincere, just saying those things to get her to sleep with him. But then he barges into the hospital and makes a point to tell her how he feels. Every convoy, every mission, gives that heightened intensity that you may never come back, so say what you mean in every moment.

I understand the events leading up to Frederic's desertion as he and Catherine disappear to Switzerland to have their baby. When I came home on R&R after about seven months in Iraq, my mother informed me that the Army had sent a letter stating they wouldn't honor the student loan repayment in my contract. I wasn't enrolled in the GI Bill program because I relinquished that benefit for the GI Bill. I deeply considered not going back to Iraq. If the military was not going to uphold their honor and duty to fulfill my contract, why would I sign my life over to them? I was convinced to return only because I didn't want my fellow medics and best friends to run those convoys, my convoys. I knew I couldn't live with the fact I'd put them in mortal danger, and I wouldn't be any better than the Army institution betraying me if I betrayed my fellow medics.

However, Frederic had that same love for Catherine as I had for my fellow medics. I understood and completely cheered him on as he deserted the military for love.

The story continues to a tragic end, which is sad yet incredibly satisfying. Frederic was able to see Catherine at least one last time right before her death. I find solace in spending last moments with the ones I am taking care of. It's truly a gift to be present at their passing from this world.

The world doesn't always have happy endings and, well, Hemingway really knows how to give an emotional "gut punch." Frederic loses Catherine and the baby. Hemingway doesn't allow a glimmer of hope. Maybe this is the metaphor of war. No one goes unscathed, no one comes home the same, and loss is inevitable.

Reading my poetry has that similar effect—jabbing your solar plexus, expelling the air from your lungs, and then sucking it from the room.

This version of *A Farewell to Arms* featured a very young Gary Cooper and, at the end, my mother called it a horrible movie in her opinion. She thinks life is hard enough without depicting such grief in a movie. Maybe she thought it was horrible because she had two miscarriages before giving birth to me, or maybe because the movie ends so abruptly with these deaths. I, however, embraced the opportunity to be sad. Rock Hudson and Jennifer Jones starred in the first film adaptation I saw, which makes it one of my favorite movies of all time, with the haunting facial expressions of Jennifer Jones, the chiseled looks of Rock Hudson, and the gut punch of emotions.

Reflecting back, I wonder if my subconscious decided one day while I was in tears that I would someday be a gut-punching writer too.

Here's a section from my poem, "Heart of the Enemy:"

Until . . .
The explosion silenced us
Your screams were deafening
Or was it mine?

I scrambled around the crater
The dust was blinding
Until I saw the blood
Desert sucks up blood
Quicker than water
I saw you!
Running away
With that cell phone
That detonator
In your little brown hand
Die You Little M-------er
You were no longer a child
With a beating heart
Sucking on cough drops
YOU ARE A THREAT
Running across your desert
Of Sand that rakes my skin
Much like your existence
Rapes my idealism

Did other old movies, like this one I watched on my parents' couch, design my inspiration, and my life experiences write the story for me?

When I was little, I saw the world like an old movie, in dramatic scenes and pictures. I have always been an absolute lover of old movies, specifically black and white. As if the black and white vision gave me the opportunity to practice my creativity in color. From the moments when the main actress enters the room with the fuzzy lens of the camera set, the dress is flowing and, in my head, I am assigning a color, if it is more of the white, I think pink; if it is darker, maybe purple or red; the magic of black and white is the sparkle of the evening gowns never let you down no matter what the color I imagined.

But *A Farewell to Arms* features no beautiful gowns, just uniforms, which it adds to the starkness of their environment. Black and white simplify the extreme measures of war. Effects today with CGI technology don't guarantee color improves war footage. When I see Iraq in my mind, the whole world is tan except my displaced green ambulance.

That scene reminds me that the Army didn't care if we had proper equipment. We are just warm bodies to fill the vacancies that war creates from the youth of our country.

I wonder if other military members realize this sad reality today and if Frederic faced the same. Maybe this was one of his less-explicit reasons for deserting. In the scene where Frederic is traveling back from the front line and hiking back with the civilians fleeing, the horrors of collateral damage are stark: a bomb lands and kills many women and children. The militaries of the world might find this acceptable in war. However, as a soldier, I found the killing of innocent lives quite unbearable, and a source of my post-traumatic stress disorder.

My writing workshops for veterans in Pennsylvania are spaces of non-judgment, kindness, and compassion. As the facilitator, I ensure that everyone adheres to these agreements. While the women and men share their writing with the group, I encourage them to hold the space and actively listen, to respond to others, or read from their work without waiting their turn. This active listening slows racing and intrusive thoughts. The groups cultivate camaraderie, and they secure a space to write and speak of their experiences, releasing them from their minds and bodies. Child suicide bombers, killing someone, rape, childhood, emotional, sexual, and physical abuse fill the workshop content. Over time, I have learned not to internalize others' pain.

After being immersed in war and trauma, going to see my parents is a huge relief. It's the downtime I have trouble creating in my own life, so I fly to Florida to find myself again, or at the very least start practicing self-care that I misplaced somewhere between my parent's house in Orlando and my home in New York City. When I watched the movie, *Farewell to Arms*, this time with them, I was, as usual, strangely both nostalgic and analytical. Nostalgic because, when I wanted to soothe myself as a kid growing up, I would watch a movie with them. Even if they disagreed about every choice I made in life, my dad and I could sit and discuss a movie. We analyzed its plot, the characters, and the accuracy of the film—he would love to show how xyz couldn't happen as filmed. I also like to analyze my parent's behavior as I see myself in both–my sarcasm, my kindness, my ability to laugh through the struggle, my overt seriousness and my silliness too. It's all there right in front of me playing out like a scene in a movie about the retirees and their combat veteran daughter who visits when she can't stand to talk, listen or discuss another trauma in a workshop or on her "downtime."

I joined the Army after getting by on multiple university campuses. I was convinced that if I tried enough colleges, I would eventually find one that stuck. What stuck was the student loan debt, which the Army promised to pay off. That didn't happen, but that's another story for another day.

My Army recruiter also said women didn't go to war. Another lie.

When I returned home from the Army, I wallowed in guilt, shame, whiskey, vodka, and needles. The company of my dogs and old movies provided hope and soothed me with their costumes, songs, and beautifully written dialogues. My times of darkness became the fuel of my poetry, no longer about rainbows and raindrops.

Ironically, I hate modern war movies. They are either inaccurate, exaggerated, or triggering to me. However, a classic black and white movie such as *A Farewell to Arms* or *Casablanca*, I will watch over and over, relishing in my feeling of relief as my creativity paints the costumes in my mind and listens to the music of that era.

Now, I write and perform poetry of the scenes from my life, as if writing through the black and white lens of my trauma or life experience. Ironically, too, I find myself grateful for the war, the drugs and poetry for leading me to the path I walk every day as a writer, lover, and facilitator guiding other veterans to find their path, their soothing, their happiness through the expression of story.

Jenny Pacanowski is the founder and director of Women Veterans Empowered & Thriving, a reintegration program that uses writing and performance to empower experiences and facilitate skills to thrive in daily life. WVE&T provides the connection, empowerment and programming for all eras, discharge statuses and any time served. While Jenny was in the Army, she deployed to Iraq in 2004 as a combat medic and provided medical support for convoys with the Marines, Air Force, and the Army. She also did shifts in the Navy medical hospital. In Germany, she was part of a medical evacuation company.

A Story of Regeneration: Ernest Hemingway's "Big Two-Hearted River"

BRANDON CARO, veteran

When I left the Navy in April 2009, I didn't have a plan. But I was interested in writing. Funds for the New GI Bill, which had been signed into law the previous year by President George W. Bush, were to be appropriated in time for the start of the school year that Autumn. I'd applied and been accepted to Texas State University in San Marcos, alma mater of President Lyndon Johnson.

I chose to be an English major. Up to that point I had never attempted to write anything serious, though I had thoroughly enjoyed working through writing tasks assigned to me in high school and later at community college in Norwalk, Connecticut. In my first semester at Texas State, I took a survey class in American Literature from the last quarter of the nineteenth century through the middle of the twentieth with Professor Allan Chavkin who remains a friend to this day.

One of the first readings assigned to us was Hemingway's "Big Two-Hearted River Parts I and II." I remember reading it for the first time on my laptop in my apartment in Austin, TX. With its fine detail, I'd never encountered anything like it.

The story follows Nick Adams, a quasi-autobiographical construction of Hemingway who appears in many of his short stories, as he takes a train ride deep into a wooded forest that bears the marks of a devastating fire that raged through the area sometime before his arrival. Nick disembarks, wandering through the woods in search of a good spot to fish. When he notices an insect with burn marks on his back crawling along the forest floor, he wonders how long it will be before the burn marks disappear:

> As he smoked his legs stretched out in front of him, he noticed a grasshopper walk along the ground and up onto his woolen sock. The grasshopper was black. As he had walked along the road, climbing, he had started grasshoppers from the dust. They were all black They were not the big grasshoppers with yellow and black or red and black wings whirring out

from their black wing sheathing as they fly up. These were just ordinary hoppers, but all a sooty black in color. Nick had wondered about them as he walked without really thinking about them. Now, as he watched the black hopper that was nibbling at the wool of his sock with its fourway lip he realized that they had all turned black from living in the burned-over land. He realized that the fire must have come the year before, but the grasshoppers were all black now. He wondered how long they would stay that way.

Carefully he reached his hand down and took hold of the hopper by the wings. He turned him up, all his legs walking in the air, and looked at his jointed belly. Yes, it was black too, iridescent where the back and head were dusty.

"Go on, hopper," Nick said, speaking out loud for the first time. "Fly away somewhere."[1]

I came to understand that the grasshoppers in the forest represent Hemingway and all survivors of the First World War, and the fire represents the conflict itself. Forest fires are naturally occurring phenomena, and trees and other flora can regenerate after a destructive fire. But if the fire is too intense, it might mean the end of a forest.

And that is where Nick Adams, and by extension, Hemingway, found themselves—waiting for their souls to regenerate after a prolonged traumatic, violent experience. And, to a lesser extent, that is where I found myself. Some of the things I saw during my yearlong deployment to Afghanistan caused me to reprogram my understanding of the world.

Loss of innocence is a common theme in war literature. It's easy to understand why. Having one's life threatened and engaging in confrontations where death or serious injury are the ultimate stakes builds character for those who survive. But operating in an environment where the prospect of death is a real and constant threat—one that places enormous stress on the nervous system—can cause irreversible damage to the psyche and soul. In the aftermath of my deployment, I discovered a newfound appreciation for security and stability; aspects of Western life which I'd, up to that point, taken for granted. I could never look at the world in the same way once I'd seen what happens when cooperation between groups breaks down.

These and other ideas I laced into my debut novel, *Old Silk Road*.[2] The story follows a motley crew of American advisors to the Afghan National Army as they travel down a long, unending road which leads them down a switchback with nine levels, an allusion to Dante's nine circles of hell. At the nadir of this descent, the soldiers learn things about the war—perhaps about all wars—that they wish they could forget.

Brandon Caro was a Navy corpsman who deployed to Afghanistan in 2006-2007. He's the author of the novel *Old Silk Road*, which is based loosely on his experiences therein. His nonfiction work has appeared in the *New York Times,* the *Daily News*, the *Daily Beast* among other places. He resides in Austin, Texas.

Notes

[1] Ernest Hemingway, "Big Two-Hearted River I," in *The Nick Adams Stories* (New York, NY: Scribner, 2003), pp. 177-199, 180.

[2] Brandon Caro, *Old Silk Road: A Novel* (New York, NY: Post Hill Press, 2015).

What the Mountains Hold: A Writer's Trek
Through the Dolomites of Mark Helprin's WWI Italy

SHANNON HUFFMAN POLSON, veteran

So like this stone are my tears unseen
Through living is death atoned [1]

–Giuseppe Ungaretti, Italian infantry WWI and poet

In the middle of September, in the mountains of Italy close to where the character Alessandro Giuliani lived and fought in Mark Helprin's novel *A Soldier of the Great War*, my husband and I headed out in rain gear while snow fell. At a local's recommendation, due to the unexpected heavy snow, we had cut out the first half of our planned day walking toward Lagazuoi and had to put aside hopes of travel along the via ferrata, or "iron ways" of more exposed and technical travel across the rock. Billows of white hid the mountain, and we set our poles and boots on icy trails ascending steep and rocky trails criss-crossed by the roots of pine and fir slick with snow.

The first few miles we climbed through deep forest. Finally, we broke out above the tree line close to the Cinque Torri, or Five Towers, a dramatic formation of rock made up of towers and a tumble of boulders. The wind picked up with nothing to stop its wildness, and we zipped Gore-tex over our chins for the last stretch of hiking before ducking into the Rifugio Scoiattoli that appeared through the clouds.

"Prego?" asked the woman behind the counter.

"Deu cappuccino, per favore," I asked, deploying the full measure of my Italian.

"Si," she said, and as we walked to a table, I heard the whirring of coffee beans grinding and the welcome whoosh of foaming milk.

A STORIED HISTORY

I'd heard that I could use a little high school German while in the Dolomites given the still strong Austrian influence, but so far that hadn't been the case. We'd find out later from a shuttle driver that two of the five valleys of the Dolomites tended toward the German language, two toward Italian and one toward neither; the ancient Ladin influence was strong throughout. The Valle del Boîte, where we had begun our trek, favored the Italian, as did the place where we'd begun, the town of Cortina d'Ampezzo.

Cortina is known for hosting the 1956 Olympics (WWII put an end to its scheduled Olympics in 1944). Our shuttle driver insisted that the area had been well known before then for its beauty and had become a popular place for wealthy visitors and royalty in the 19th and early 20th centuries. He and his wife were of the older Ladin group, which, he explained, claims a written history going back over five-hundred years. The Ladins later organized into a community owning 80% of the land in the Dolomite area. The group recognizes membership exclusively through male descendants.

The civilization of the Valle del Boîte, however, extends back into prehistory, as evidenced by the discovery in 1987 of a primitive tomb at Mondeval de Sora, as well as the discovery of 5,300-year-old Otzi in 1991 in nearby Bolzano. More recent history slowly reveals itself, too, as Italy's glacier's melt to reveal the bodies of soldiers from WWI; the remains of three soldiers believed to be from the Austrian-Hungarian Army were discovered on the Presena Glacier in 2012.

DISCOVERING THE DOLOMITES OF WWI

We found our way to a table by the far window of Rifugio Scoiatolli and pulled off our coats. In the warmth, the blood rushed back into our fingers and faces with sharp pricks of pain. But for one table across the room occupied by two other synthetic-clad trekkers, the tables were empty. The woman who had taken our order delivered shallow round coffee mugs on saucers with tiny spoons and sugar, and we gratefully sipped the warm foam from the top of the cappuccinos. Outside, the dark shapes of trekkers moved against the whiteness, the snow accumulating to several inches on the deck outside the window. Then, the snow still falling, the clouds thinned to reveal the Cinque Torri. The towers stood out against the clouds like sentries, their dark and precipitous shapes beautiful in their extremity.

We finished our cappuccinos. Donning our jackets and hats, we headed out to circle the Cinque Torri. The earliest part of our hike we had been forced to abort due to early season snow would have passed the front lines between the Austrians and Italians from the first World War, replete with long tunnels and caves. Shortly after beginning the circuit of the Cinque Torri, however, began a network of well memorialized and maintained remnants of WWI fighting. These comprised the second line, looking down on the intense fighting below, spotting for artillery.

A few years prior I walked with my family in the trenches of the Somme. This is the place I most associate with WWI, the lines and lines of trenches in the mud where Wilfred Owen penned his "Dulce et Decorum Est," and John McRae "In Flanders Fields," and where small cemeteries lie littered throughout the French farmers' fields, bodies buried where they fell. I had never considered the war in the Dolomites, and first saw it referenced by reading about the world-class French aviatrix and athlete, Marie Marvingt, who was not permitted to use her considerable aviation skills in the war and instead is believed to have disguised herself as a man to fight at the western front until she was discovered. When she was forced to depart, she is believed to have joined in military operations with the 3rd Regiment of Alpine Troops in the Dolomites.

No place for similar small cemeteries exists in the rocky Dolomites. The recovered bones, both Austrian and Italian, lie in a tower in Cortina which did its best to remain neutral during the fighting. Before WWI, it was a part of Austria, but according to one placard mounted in the trenches, the mayor advised residents to welcome Italian soldiers to the city and not to meet them with hostility. The fighting would be left to soldiers. Other sources suggest the military-aged men of Cortina made an unsuccessful attempt to fight the Italians.

Along 400 miles of front in the Dolomites, broken by the severe terrain, 689,000 Italians died (out of a population of 35m) and 400,000 Austrians between June of 1915 and October of 1917, though the sparse record keeping makes exactitude impossible. Hoping for territorial gains, Italy was a late entrant into the war. At the end of the fighting, Italy was awarded the Dolomites region, soaked in Italian blood. According to Iraq veteran Brian Mockenhaupt writing for the *Smithsonian*, it was fighting "like none the world had ever seen, or has seen since."[2] Over 100 years ago, E. Alexander Powell wrote that "On no front, not on the sun-scorched plains of Mesopotamia, nor in the frozen Mazurian marshes, nor in the blood-soaked mud of Flanders, does the fighting man lead so arduous an existence as up here on the roof of the world."[3]

This is the terrain—or not far from it—best known through Ernest Hemingway's *Farewell to Arms,* and I was excited to read a newer, if also fictional, account of the war in this place. I'd begun reading *A Soldier of the Great War* before arriving in Italy, a novel following the extraordinary military career of Alessandro Giuliani through remarkable and horrific experiences in WWI Italy and beyond, including two tours in the Dolomites mountains. He spent his childhood in an affluent family climbing and skiing in the Alps. Unlike most of the soldiers who headed north, he knew as a child how the "mountains acted upon them and their spirits were calmed and enlarged."[4]

Giuliani initially serves to defend what he calls "the Bell Tower" along the Isonzo front, the Italian name for Slovenia's Soca valley where most of the Italian casualties occurred, and also where Hemingway's *Farewell to Arms* takes place, which is north and east of Cortina. It might be any piece of the Dolomite line, as the shapes of the huge formations all lend themselves to similar descriptions. The details of trenches and facilities are grim, though the soldiers take liberties and risk sneaking out of their lines to swim in the river at night. At one point, a soldier holds a helmet on a stick above the trench, and it is immediately shot by the Austrians. One has the sense of misery and constant engagement at close quarters, despite the beautiful setting.

In his letter home, the character Giuliani writes:

It's too tense here, and everyone is too unhappy . . . we are constantly expecting an Australian to come from nowhere, throw a grenade, fire some shots, and bayonet a poor idiot coming out of the latrine. This kind of thing makes you tense 24 hours a day. So do the shells. On average, 8-10 a week hit the Bell Tower and you never know when they're coming . . . Another sense of tension here is that we have no privacy . . . I don't care about our claims on the Alto Adige, so I'm fighting for nothing, but so is everyone and that's not the point. A nightmare has no justification, but you try your best to last through it, even if that means playing by the rules.[5]

Much later, Giuliani is sent back to the mountains where he is told "in the mountains the trench works are shallow and inadequate."[6] As he reports in his letter to his family, when morale sagged, the inept generals of the Italian Army resorted to the ancient Roman practice of decimation, executing one man out of every ten at random.

THE THIRD ENEMY

Historian Mark Thompson describes the weather and terrain in the Dolomites as a third enemy, accountable for more deaths than the bullets of either side. The Italians called the fighting in the Dolomites "il fronte verticale." On December 13, 1916, what became known as White Friday, an avalanche covered over 10,000 soldiers. Both armies used huge amounts of explosives to not only create tunnels for movement, but to use the rock itself as ammunition. The challenges of hauling ammunition and artillery pieces up sheer cliffs and across snowfields and glaciers resulted in casualties as well. For two years, soldiers died, and no gains were made by either side.

The maze of trenches cut and maintained in the rock around Cinque Torri leads at times to caves cut into or under the rock where mannequins demonstrate what troops might have been doing, and one expects tunnels are here as there are across the valley. Snow blows into these trenches and walking there we find only a small escape from the wind. Soldiers would have had little respite. Our trip is technically a summer trek, and we have not brought gloves, which we regret. Soldiers who were stationed in this piece of high mountain rock would have experienced much worse, temperatures 40-50 degrees below zero in winter and howling sustained mountain winds. I stand in one bunker and look out beneath the low cover at the valley across, imagining incoming artillery fire, imagining that there was no rifugio in which to dry my clothes and crawl into bed, imagining the terror of the enemy, of the terrain and of the leadership, imagining the unimaginable physical and psychological fatigue and the resignation to death.

A short hike beyond lay our refugio, named Averau for the peak just above it. Here the Italian soldiers could take refuge themselves, in quickly constructed dormitories. The Italian army set a wooden staircase up the side of Averau to put in an observation point for three soldiers, to observe Austrian movement, take photographs and illuminate attacks, and adjust artillery fire by using Morse Code with a large spotlight. No trees grow this high. All movement on the steep rock must have been at night.

TRAVELING SACRED GROUND

Around the Cinque Torri, a cross still stands in the rock off one of the trenches, a reminder that this scene, this landscape, is not to be traversed lightly. How do you travel through sacred ground? Pain and tragedy live alongside the harsh beauty of the light stone. My husband has gone ahead down the trail, and through the wind I hear the click of his trekking poles against the rock. I look up at the rock towers above. As a native Alaskan, I've always gone to the mountains for solace, for peace, for challenge. I love the roughness of steep terrain. The soldiers who came here, though, would have come from the flatlands far below. Except for the character in *A Soldier of the Great War*, most would have thought these mountains, as did H.G. Wells who visited in 1916, "grim and wicked."[7]

When we embarked for the Dolomites, we came for the peace and vitality of the mountains, the clarity that comes from pushing our bodies in high rocky places, but these trenches give me pause. I too have worn a uniform, followed orders, accepted the rigor and discomfort of circumstance, if at a different time and in a different way. In the midst of the blowing snow come whispers of the valley's history across all of time, adding gravitas to the already stark requirements of mountain travel. This land is sacred ground. If not soldiers of my country, they were soldiers here nonetheless. All those who take an oath or submit their lives to service share a bond, across generations and geographies. I know the feelings of despair, elation, resignation and fear. I wonder if I might honor them if not by memory, by imagination.

Our trek has just begun. Days of trekking lie ahead of us, in and out of other valleys, over other passes and high places swept by histories and wind. As we've traversed the Cinque Torri, still visible through the cloud, I wonder if those soldiers one hundred years ago might have paused amidst the horror, and while they watched for enemy saw too the light across the faces of the mountains and if, the way I know it, they found a sense of peace. Helprin shares his hope that this is true; when Giuliani is sent back to the lines, where he and his fellow soldiers assume they will die, his commanding officer delivers a resigned message to the troops. "Should you not make it back to the table in the piazza, you will have died in the best place for dying the world has ever known. What I mean is, here, you're practically at the gates."[8]

The Italian poet Giuseppe Ugaretti found glimpses too, among the carnage:

Suddenly
high above the rubble the pure
wonder
of the infinite appears[9]

Shannon Huffman Polson is an American author, speaker, and entrepreneur. She is the author of *The Grit Factor: Courage, Resilience and Leadership in the Most Male Dominated Organization in the World* (Harvard Business Review Press, 2020), and *North of Hope: A Daughter's Arctic Journey* (Zondervan, 2013). Polson served as an Army aviation officer in the 1990s and was one of the first women to fly the Apache attack helicopter.

Notes

[1] Giuseppe Ungaretti, "I am a Creature. Valloncello of Peak 4, August 5, 1916" in *A Major Selection of the Poetry of Giuseppe Ungaretti,* trans. Diego Basianutti (Toronto: Exile Editions, 1997), 69.

[2] Brian Mockenhaupt, "The Most Treacherous Battle of World War I Took Place in the Italian Mountains," Smithsonian.com (Smithsonian Institution, June 1, 2016), https://www.smithsonianmag.com/history/most-treacherous-battle-world-war-i-italian-mountains-180959076/.

[3] E. Alexander Powell, *Italy at War* (New York: Scribner, 1918), 101.

[4] Marc Helprin, *A Soldier of the Great War* (New York: Houghton Mifflin Harcourt Publishing Company, 1991). chap.1, Kindle.

[5] Marc Helprin, *A Soldier,* chap. 4, Kindle.

[6] Helprin,chap. 6, Kindle.

[7] HG Wells, *War and the Future. Italy, France and Britain at War* (London:Cassell and Company, 1917.)

[8] Helprin, chap. 7, Kindle.

[9] "Vanity, Vallone, August, 19, 1917" in *A Major Selection of the Poetry of Giuseppe Ungaretti,* trans. Diego Bastianutti (Toronto: Exile Editions, 1997), 121.

part two

PAST AND PRESENT: BRIDGING THE WWI MILITARY-CIVILIAN DIVIDE

More Genteel Than Grim:
Letters Home from WWI

DAVID CHRISINGER

From 2014-2017 I taught a seminar at the University of Wisconsin-Stevens Point for student veterans on the history of American veterans coming home from war. In the fall of 2015, the students in my class analyzed hundreds of letters written by soldiers fighting on the Western Front during WWI who had grown up in Stevens Point. As they read each letter, I asked them to highlight passages that struck a chord with them, that reminded them of their own experiences in Iraq or Afghanistan or elsewhere.

What they found surprised them: the letters were more genteel than grim.

They were surprised, though perhaps they shouldn't have been, that most of the letters came from young and confident American doughboys, not the con-flicted old men my students think of when picturing a WWI veteran. One soldier wrote in March 1918 that he was "perfectly satisfied, feels great, sleeps like a log, and wouldn't come back until we win the war for anything." Another wrote that "this sure is the life. Why the hell didn't I get into the Mexican war? You can bet I'll never miss any more wars if I know they are in progress."

In their letters, the soldiers are sturdier than their enemies because they are also nobler, more courageous, and more compassionate. They are more human than the "Boche" but also less corruptible. One soldier wrote that he was "talking with an old lady (refugee) Sunday who had her foot shot off by a 'Boche' because she refused to feed some soldiers. Another was telling me about her two grand-children being bayoneted by them—three and six year olds." He finished his let-ter sarcastically, "They love the German nation!" Another soldier talked about souvenir hunting and wrote home that he had "collected a few." Most notewor-thy was "one of those belts with 'Gott mit Uns' [God is with us] buckle." The soldier was offended by such irony. "They sure do have a pile of nerve to have that on their clothing," he wrote.

The letters also reveal a desire to entertain with anecdotes, to make the Western Front vivid where they can, but not too grim. Their toned-down, respectful language is unmistakable. One soldier, for example, wrote about being shelled at the front: "We were supposed to be relieved that evening but the fun started too early so we were right there. And oh boy, when those guns let loose with their roar I thought sure that I was in that place where a snowball is so short lived."

Many of the letters are written from the point of view of American heroes who understand the gravity of their endeavors but who are too humble to take much credit. They play small parts in a great drama unfolding around them. One soldier, a medic, writes this about the first battle he witnessed:

I wish I could describe that first advance of ours to you. Right out in the open, up a hill, every man with a bayonet and all keeping a straight line. One could see our creeping barrage, perhaps a hundred yards ahead of them. Right up they went with never a one faltering. They reached the woods alright, with fewer men than they started with, but they kept right on and on. It was wonderful and gruesome, too, but it fascinated a person. Now our work comes in, getting those men to the dressing station. It's nothing compared to the infantrymen's job, but we worked until we absolutely dropped in our tracks.

Perhaps the strongest parallel my students noted to their experiences dealt with the fundamental paradox of military service—while it gives so much in the way of training, experience, and friendship, it also takes much away. As relatively new veterans, this is something most of my students hadn't yet made sense of, much like the soldier who wrote this to his parents in December 1918: "Where do you people get the idea I was lucky in getting over here? I can't see it. I think the fellows back home were lucky, but then I would not have missed my experiences for the world."

In the few years I taught student veterans, I saw again and again how this history powerfully impacted them. When they discover that others have walked their same path, they get much needed perspective. They see that their challenges are not as unique as they once believed, their challenges are not insurmountable, and that every generation of American veterans has had to face eerily similar challenges, both on the battlefield and once they return home.

After we finished analyzing the last batch of letters, a quiet hush fell over the classroom. For a few moments I could see the students processing the material, each in their own way. After about a minute of silence, one of the students looked toward the front of the room, locked eyes with me, and leaned back in his chair. "Times change," he said, "but all the shit that happens in war sure hasn't."

David Chrisinger is the director of the Harris Writing Program at the University of Chicago, where he teaches policy design and communication. He wrote *Public Policy Writing That Matters* and serves as the Director of Writing Seminars for the *War Horse*. His book based on his teaching was also published by Johns Hopkins University Press, titled *Stories Are What Save Us: A Survivor's Guide to Writing About Trauma*. His research and writing topics include education; retirement security; social policy; and issues related to military veterans' transitions back to civilian life. In addition to the years he taught student veterans at the University of Wisconsin-Stevens Point, David also taught public policy writing to graduate students in the Public Policy Program in the Bloomberg School of Public Health at Johns Hopkins University. He is the author of *The Soldier's Truth: Ernie Pyle & the Story of World War II*, forthcoming Penguin Books spring 2023.

I Never Saw Him Drowning:
Great-Uncle Charlie and the Great War

PHILIP METRES, PhD

Leaning over my desk in January 1991, news coverage of the Gulf War droning in the background, I read for the first time the opening lines of Wilfred Owen's "Dulce et Decorum Est:" "Bent double, like old beggars under sacks,/ Knock-kneed, coughing like hags,/we cursed through sludge."[1] I felt transported, imagining the weight these young soldiers bore in the strange hellscapes of the front. I was a junior at Holy Cross College. All semester, my English professor, Mike True, guided our class through "The Nonviolent Tradition in Literature," while on television, military leaders touted missile-eye images of "smart bombs" and "surgical strikes." It was surreal to encounter such polar views of war, knowing that each was a partial picture of what American poet Walt Whitman once called the "Real War," the one that would never "get in the books."[2] Yet Owen's garish images and his fierce rhetorical conclusion confirmed something that I felt deep in my gut— war was an ugly thing, destroying bodies and haunting minds. On the other side of the world, though news coverage would not show it, people died under our bombs; it would take poets and artists to slip beneath the media's redactions and censorship to imagine the horror.

Even then, I found myself drawn most, not to the lurid language and angry retort of "Dulce," but to its dreamlike center when the soldiers find themselves attacked by chemical gas, and one man fails to adjust his mask in time:

> Dim through the misty panes and thick green light,
> As under a green sea, I saw him drowning
> In all my dreams before my helpless sight,
> He plunges at me, guttering, choking, drowning.[3]

In contrast to much of the poem, the language here is starkly plain and subtly recursive, filled with internal rhyme and assonance. Words flicker and return: green and green, drowning and drowning, and the piling on of gerunds creates the effect of a flashback, of a past imposing itself on—even devouring—the present. Each night, the speaker drowns in the memory of his comrade's drowning. In WWI, after Owen's dear friend was killed in a bomb attack, the poet suffered from acute mental distress and wound up at Craiglockhart Hospital in 1917 for treatment. They called it "shell shock" then, what we might diagnose now as Post-Traumatic Stress Disorder. Though some treatments included electric shock therapy, Owen's doctor recommended writing.

It was no surprise that I linked Owen's depiction of WWI to the Gulf War; my father is a veteran and my mother a pacifist. After joining the Navy through the ROTC program in college, my father served during the Vietnam War, an advisor on a Vietnamese patrol gunboat who survived Tet. Upon return home, he taught counterinsurgency and later produced a dissertation on the emotional impact of war on families of soldiers who were prisoners of war or missing in action (POW/MIA). He went on to become a clinical psychotherapist.

In recent years, it has been his great passion to work with veterans making the long emotional return from all of our wars, but back then, when I was young, he may have been making the long return himself. I want to say that he still carried the war with him, but it's probably more accurate to say that I didn't know where the war ended, and my father began. A deeply loving and powerful man, he was nonetheless prone to suddenly foul moods that would descend upon our whole house, causing us to scurry around for whatever mask might protect us. Was this the war, I wondered, or something else? I never could be sure.

My mother abhorred violence—both real and represented—and forbade guns (real and play) in the house. She was chagrined when my father pointed outside one day, where I had picked up a stick in our yard and pretended to shoot my friend with it. Part of her pacifism grew out of a story that she'd been carrying for years, one that goes all the way back to WWI.

Her mother, Grandma Sheila, was just six years old when her big brother Charlie, age twenty-two, went off to fight. It was 1917, the same year that Owen was convalescing at Craiglockhart. My grandmother must have looked up to him, literally and figuratively, as he stood in the doorway in his trim wool uniform and its gleaming bronze buttons, saying his goodbyes.

About a year later, when the war was supposed to be ending, her parents sat her down and gave her the news. Charlie would not be coming home. He had died in the war. It must have shaken her to the core. How could this strong and beautiful man be gone? She would always look up to him, and imagine him in heaven, looking down at her, protecting her.

Fifty years later, precisely the same year that my father was in Vietnam, Sheila opened a curious letter from the Veterans Administration. It was a notice that her brother Charlie had recently passed away in a VA hospital.

I imagine her falling into a chair, rereading the letter. Her brother had been alive, and she hadn't known. She must have been beside herself with grief because everyone had pretended he was dead. Since her parents had passed away, she could never ask them why. Why had they hidden the truth? Had anyone from the family gone to visit him? If so, why was she robbed of the chance to do that same?

The reasons have been lost to the great silence of time.

T.S. Eliot wrote: "Human kind cannot bear very much reality."[4] We don't like to hear that war undoes so many lives, but if we don't listen to what its military and civilian veterans tell us, we risk repeating the same mistakes.

Because we never know whether the family stories we're told are true, I decided to do my own research and combed through an ancestry database to find out what I could about Uncle Charlie. I found his registration card for WWI and census records from 1920 and 1930, in which he was listed, in careful cursive, as "patient" in two different hospitals.

Many names appeared alongside Charlie's, men who had disappeared from their pre-war lives. How many Charlies were pronounced dead before their time, silenced out of family shame, confusion, or despair?

Great-Uncle Charlie is one of the many reasons that I've devoted a good deal of my life work—my poetry, essays, and scholarship, not to mention my teaching—to resisting the glamour of war and promoting peacebuilding and conflict transformation. From *Sand Opera* to my translations of poets from enemy countries, from my book on poets and the peace movement to my courses on Israel-Palestine and Northern Ireland, I've tried to find ways of understanding the geography of violence and pointing out the paths to peace, justice, and reconciliation.

This is all I know about Charlie Fitzpatrick, my great-uncle. According to his military registration card, he was 5-foot-7 and slender. His eyes were blue, and his hair was brown. Prior to his service, he'd worked at Western Union. He'd tried to claim exemption from the draft because he had nasal trouble, though its cause is not listed. He must have had trouble breathing, a physical weakness I may have inherited. I was suddenly cast back to Owen's "Dulce et Decorum Est" and the images of a man drowning, sputtering in the green sea of a chemical attack. After fifty years in a Veterans Administration hospital, Charlie died in 1967, just as images of my father's war had begun to flood the nightly news.

In 2015, I published the following poem about my uncle and father, "The Other Side:"

In the First World Peace, Uncle Charley found his mind outside a country asylum, wound home to meet the sister who always knew he'd survived. Uncle Dom always spoke of the Second World Peace, how he once gave away a Purple Heart, which never reminded him of anything. And the marchers marched backward, never showing their backs, unlocked and unloaded every rifle they carried, breaking ranks in every direction. And the smoke and ash turned back into bone-clothed skin. In Saigon, my father saw fireworks, wondered when peace would end. If only we trimmed our peace budget, the President says, we might have something to spend on the latest bionic technology, to erase the scars of peace, since men keep sprouting limbs, and their shrapnel-free faces have a symmetry you only see in times like these.[5]

Philip Metres is the author of ten books, including *Shrapnel Maps* (2020), *The Sound of Listening: Poetry as Refuge and Resistance* (2018), *Pictures at an Exhibition* (2016), *Sand Opera* (2015), and others. His work has garnered the Guggenheim Fellowship, the Lannan Fellowship, two NEAs, six Ohio Arts Council Grants, the Hunt Prize, the Adrienne Rich Award, three Arab American Book Awards, the Watson Fellowship, the Lyric Poetry Prize, and the Cleveland Arts Prize. He is professor of English and director of the Peace, Justice, and Human Rights program at John Carroll University

Notes

[1] Wilfred Owen, "'Dulce Et Decorum Est" *Poetry Foundation* (Poetry Foundation), accessed January 12, 2022, https://www.poetryfoundation.org/poems/46560/dulce-et-decorum-est.

[2] Walt Whitman, "The Real War Will Never Get in the Books," in Specimen Days & Collect (Philadelphia: David McKay, 1882), 80.

[3] Owen, "Dulce."

[4] T.S. Eliot. "The Four Quartets." Collected Poems: 1909-1962. (San Diego: Harcourt Brace Jovanovich), 176,

[5] Philip Metres. "The Other Side." Prairie Schooner 89, no. 1 (2015): 126-126. doi:10.1353/psg.2015.0099.

Fictions of Rehabilitation

MARK WHALAN, PhD

My recent book examines the Homefront in the U.S. in WWI, and specifically how American literature engaged with the fierce debates that roiled society over the rights and responsibilities of citizenship in wartime, and also over what the state was empowered to do in moments of national emergency. The U.S. state expanded into new areas of American life in the war—through conscription, the nationalization of the railways, the Food Administration, and the mass manufacture of propaganda, for example. Authors considered the impact of this on everyday life experience, and how they could imagine the new communities and identities that this expanded state power had produced.

Government-run healthcare became an interesting arena, as WWI saw a complete transformation in the ideas and institutions the government deployed to assist wounded veterans. These changes revolutionized customary ideas about disability, and about male citizenship. These changes were taken up and contested by authors of the era, some of whom were wounded veterans themselves.

My analysis drew heavily on two recent works of history—Beth Linker's *War's Waste: Rehabilitation in World War One America* and John M. Kinder's *Paying with their Bodies: American War and the Problem of the Wounded Veteran.* Linker and Kinder discuss the War Risk Insurance Act, or WRIA, which passed in 1917 and broke decisively with the pension system used to support Civil War veterans. By 1917, it was a byword for bloated government spending and political clientelism.

It's hard to believe now, but before WWI, the U.S. was infamous for its lavish, and even bloated, welfare state, a reputation due to expenditure on Civil War pensions. By 1915, the U.S. had spent more on Civil War pensions than it had on prosecuting the war. As a consequence, it became a target for progressive reformers keen to reduce budgetary "waste" and eliminate political cronyism.

In anticipating a new generation of wounded veterans, the WRIA sought to reform this system. It aimed to provide maximalist medical cures to wounded veterans rather than pensioning them off, an ambition hitched to the expectation that disabled veterans should undergo medical rehabilitation treatment and then rejoin the workforce. As Kinder observes, rehabilitation was "an integrated program of physical and social reform combining orthopedics, vocational training, psychological counseling, and industrial discipline" that became the cornerstone of how the U.S. government would frame its obligations to the disabled veterans of World War One and in all subsequent wars.[1] The government undertook a massive hospital building program in support of this effort, building fifty-eight general and specialized hospitals nationwide controlled by the new VB, and also established an enduring "ethic of rehabilitation," in Linker's words, which proposed that "medicine can cure disability and, more important, solve the social problems brought about by a war that fundamentally disrupts the lives of its citizens."[2]

Yet this ethic also meant that soldiers were no longer heroic simply because they were wounded. Now, full male citizenship was achievable through a successful return to the workplace, ideally in work that supported a family and a wife who stayed at home. Retirement and subsistence on a pension—being a "home slacker," in the parlance of the time—was increasingly seen as infantilizing or feminizing. The WRIA had the government pay in full for disabled veterans' medical care for the first time, but it also made rehabilitation mandatory, refusing honorable discharges for those servicemen who rejected it.

Such sweeping changes necessitated a government media offensive. Consequently, the "ethic of rehabilitation" was loudly and diffusely promoted across the Progressive press, and especially in the kind of public information apparatus of state-funded magazines and publicity bureaus that was such a hallmark of Wilsonian Progressivism. Most interesting to me was *Carry On: A Magazine on the Reconstruction of Disabled Soldiers and Sailors,* produced by the Office of the Surgeon General and which ran for ten monthly issues, beginning in June 1918. Aimed at veterans, the general public, and medical professionals alike, it presented a mix of essays, speeches, feature articles, photographs of rehabilitated veterans, fiction, cartoons, and even comic sketches. In general, it promoted a vision of veterans enacting a smooth transition from war disability to success in the workplace and in the marital home, through the agency of rehabilitative medicine and vocational training.

Carry On published pieces by Woodrow Wilson, Theodore Roosevelt, Samuel Gompers, Charles M. Schwab, and John Galsworthy. From its March 1919 issue onward, the magazine's frontispiece was the "Creed of the Disabled:" "Once more to be useful—to see pity in the eyes of my friends replaced with commendation—to work, produce, provide, and to feel I have a place in the world—seeking no favors and given none—a MAN among MEN in spite of this physical handicap."[3]

Most of *Carry On*'s pieces echoed this creed, and rejected what it characterized as the culture of dependency fostered by the Civil War pension system. It was full of pieces on what disability scholars term "supercrips," individuals lauded for "triumphing" over disabilities to win extraordinary success. Theodore Roosevelt, perhaps the most famous American to have "overcome" a disability, was the subject of an early, lavish puff piece of this kind. In addition, the more extreme of the magazine's editorializing suggested that successful reintegration to the workplace was largely a matter of individual willpower. As one editorial writer put it, "the only hopeless cripple is a deliberate shirker."[4]

Yet just as important to the magazine as this kind of high-handed exhortation to wounded veterans about their ongoing medical obligations were its attempts at "reconstructing the public," in the words of Assistant Editor Arthur H. Samuels.[5] This broader public campaign aimed to encourage well-meaning wives and mothers not to let their wounded husbands and sons be "content to be an invalid" in the family home under their care, but to engage with the institutionalized practices of rehabilitation on offer from the government.[6] It also criticized employers for well-entrenched practices of discriminatory hiring that excluded disabled men from the industrial workplace. Accordingly, the magazine boldly imagined itself leading a wide-scale social transformation of attitudes toward war disability and male citizenship.

But even in the pages of *Carry On,* not all authors entertained such ambitions. In particular, some fiction writers were loath to abandon sentimental narratives about the potential of feminized, emotional care to restore wounded American men. Edna Ferber's short story, "Long Distance," is one example.[7] Set in an American rehabilitation hospital in England, it focuses on an incapacitated A.E.F. veteran and brawny Chicagoan who has lost his memory from shell-shock. It pokes fun at what it calls this "he-man" doing the dainty work of making tiny toy chickens in occupational therapy. In the end, he snaps back to his normal self not due to the wonders of modern rehabilitative medicine but because of a letter from his sweetheart at home.

The story, therefore, endorses well-entrenched ideas about the gendering of disability and therapy that *Carry On* elsewhere worked hard to overturn. The story mocks occupational therapy as infantilizing and effeminizing. Such condescension was in sharp contrast to the regular features of occupational therapy in the magazine, which occupied 1,200 women in American reconstruction hospitals in 1919. Similarly, *Carry On* took a stern attitude to the place of feminized "sentiment" in the practice of rehabilitation. It often characterized female compassion and sympathy as hindering wounded soldiers' motivation to return to the workforce, and as repeating the mistakes of the Civil War system, which, they asserted, encouraged pension-dependent men to languish, unproductively, in the family home. The rehabilitation program actively trained their occupational therapists to avoid sentimental attachments, and it discouraged excessive sympathy among wives and mothers of disabled veterans.

Yet writers of popular fiction were less keen to relinquish the idea that women in the home had a unique curative power, an idea which had long been a staple of sentimental fiction. Instead, they frequently characterized sympathy or emotional connection from sweethearts, mothers, wives, or fiancées as carrying unparalleled therapeutic value. "Long Distance" is one such example, but there were many others at the time, including several novels by Zane Grey and stories by Sinclair Lewis. American writers, therefore, tasked the new institutions of rehabilitation not to abandon the languages and the spaces of sentimentality, which were so deeply embedded in cultural scripts about restoring disabled veterans to full vocational and marital achievement.

Veterans' fiction also forcefully pushed back against the new ethic of rehabilitation. It often countered rehabilitative rhetoric's optimistic account of willpower and discipline triumphing over embodied damage by forcefully identifying the stubbornness of injuries which refused to heal. And, it spotlighted the most glaring absence in the rhetoric of rehabilitation, namely pain. Pain was a word rarely found within *Carry On*'s cheerful accounts of rehabilitation—indeed, there were only ten uses of the word across its entire run. And partly in consequence, pain was wielded as a rebuke to such rehabilitative rhetoric in the work of several veterans who wrote about the process of coming home. In work by William March, Ernest Hemingway, and especially Laurence Stallings, pain becomes the center of both everyday experience and the practice of citizenship.

Most interesting to me was that few of these fictions completely rejected the new rehabilitation program. Instead, they tended to urge more complex accounts of how and where care occurred, and sought more embodied, experiential accounts of what it meant to suffer injury and receive medical treatment. As such, we can see these fictions of rehabilitation not just as attempts to understand new state institutions, but also as attempts to shape them.

Mark Whalan is Robert and Eve Horn Professor of English and Head of English at the University of Oregon. He is the author of *World War One, American Literature, and the Federal State* (Cambridge University Press, 2018), as well as *American Culture in the 1910s* (Edinburgh University Press, 2010), *The Great*

War and the Culture of the New Negro (University Press of Florida, 2008) and several other books and articles about American literature and culture in the early twentieth century.

Notes

[1] John M. Kinder, *Paying with Their Bodies: American War and the Problem of the Disabled Veteran* (Chicago: The University of Chicago Press, 2016), 117.

[2] Beth Linker, *War's Waste: Rehabilitation in World War I America* (Chicago; London: University of Chicago Press, 2014), 180.

[3] Anon, "Creed of the Disabled," *Carry on. A Magazine on the Reconstruction of Disabled Soldiers and Sailors*, no. 6, (1918): i.

[4] Herbert Kaufman, "The Only Hopeless Cripple," *Carry on. A Magazine on the Reconstruction of Disabled Soldiers and Sailors* 1, no. 4, (1918): 22.

[5] Arthur H. Samuels, "Reconstructing the Public," *Carry on. A Magazine on the Reconstruction of Disabled Soldiers and Sailors* 1, no. 4, (1918) : 15–16.

[6] Alice Duer Miller, "How Can a Woman Best Help?" *Carry on. A Magazine on the Reconstruction of Disabled Soldiers and Sailors* 1, no. 1, (1918): 17–18, 17.

[7] Edna Ferber, "Long Distance." *Carry on. A Magazine on the Reconstruction of Disabled Soldiers and Sailors* 1, no. 4 (1918): 2–6.

"TWITOS":

Journalist Connects French
Youth to WWI *via Twitter*

STEPHANIE TROUILLARD

Throughout the Centennial in France, which began in November 2013, I wrote more than one hundred articles on WWI. I broached many subjects—military subjects, but also cultural, sports, and even scientific subjects. Every day, I would compose a press review on my Twitter account and give news about the Centennial. At the same time, I was also researching the *Poilus* of my family. *Poilu* is the French term for WWI soldiers. It means "hairy," which is how the French press during WWI described the haggard, and, yes, hairy soldiers with those big mustaches. I would publish the stories in my news articles or on Twitter and then go back to the field, continuing to trace the diverse ways WWI impacted my family. I discovered three of my great-uncles lost their lives and two of my great-grandfathers served. They were in different armies and regiments: the infantry, the artillery, and the navy. This family connection has allowed me to have a more intimate approach to this great story of the Great War as a journalist.

Thanks to the community that France 24, a major French media outlet, created on Twitter, I regularly received messages from the descendants of other *Poilus,* who would send me WWI documents belonging to their family. These sources are important because since they've gone unpublished, we've never seen them before. Above all, they are particularly moving because they are personal.

I was also contacted by high schoolers and middle schoolers to talk about my work on WWI. I love this exchange with students because I got to know their outlook at our country's past. Digital resources made it easier to interest them in history. They use modern tools to immerse themselves in events that took place 100 years ago. It's a gateway to encourage them to research real, non-digitized archives because, in the end, nothing is equivalent to direct, manual contact with historical documents. I've experienced very strong emotions when touching old papers with my hands, whether it be a military registry, an obituary, or a simple newspaper clipping. Nothing replaces this direct contact with the past.

Nonetheless, although it may seem a far cry from archive research, Twitter has been a great tool for reaching a wide audience, especially the youth. I often got messages from high school students who followed me and who were passionate about this period. History teachers, who often relied on my articles to talk about the Great War with their students, followed me, too. Over the course of one scholastic year, I worked with a class at the Sotteville-Lès-Rouen high school in Normandy thanks to Twitter. The students would tweet me questions about their research on *Poilus*. I also came to the class in person to meet the students and explain my work to them. It made me see we can study the past with the tools of today.

WWI is not old fashioned. Just look at the success of the hit video game *Battlefield* that takes place during the war!

Admittedly, I was astonished by the over 12,000 people on my Twitter account. It's easy to think that history's not a trendy subject and it only attracts older people, but I saw a diverse community created around commemorating the Centennial.

Now, when I go out in the field as a reporter for France 24, I don't hesitate to interact with Twitter users, no matter what I'm covering. During the Centennial Battle of Verdun in 2016, I asked the "Twittos" (Twitter users) if they had soldiers from their families who had participated. Many sent me moving messages with pictures of their ancestors. And, when I would go to cemeteries, people ask me to take a picture of their family's grave. I would send it to them, and we would discuss such or such battle. I really liked this interactive and instantaneous side.

In 1914, the Internet didn't exist, obviously. Neither did Twitter or Facebook. Would social networks have influenced the unfolding of WWI? Today we're flooded with images of conflicts in Iraq or Syria. Every day on our Twitter or Facebook feed, we see photos or videos of fights or injured children. Social networks are moving Internet users, but it's regrettable they don't have the power to stop wars.

Stéphanie Trouillard is from Nantes, France and is a journalist for France 24. Over the past few years, she has specialized in the history of the First and Second World War. In addition to creating the most popular WWI Twitter feed in France during the Centennial, she also produced a web documentary in 2017 entitled, "If I Ever Come Back," about a Jewish Parisian schoolgirl's letters she wrote before she was deported and killed in Auschwitz. She also has published a book called *My Shadow Uncle,* which tells the story of her investigation into the traces of her great-uncle killed in July 1944, in a Breton maquis during WWII. Today, she is invited all over France and throughout Europe to talk about her various projects about both world wars.

part three

WWI AND WOMEN TOO: FIGHTERS, NURSES, WRITERS

WWI *Navy Yeoman (F) First Class Marjory Stoneman Douglas*: Writing Advice, Writing Life

JERRI BELL, veteran

Her short stories weren't bad, Marjory Stoneman Douglas said in her autobiography, *Voice of the River:*

> [...]but they weren't the newfangled concise dramatic short stories of the Hemingway school. Though many of the men who'd served in the First World War had been prepared for that kind of Hemingway writing, I was not a part of it. I didn't subscribe to the Hemingway thinking. I was more or less tied into the mainstream from which Hemingway was estranged. I couldn't write in that bare, stark way in which a story begins like a slap in the face.[1]

This may be one of the reasons that Douglas is not counted among the "war writers" of WWI despite a distinguished literary record, which include work featuring characters, settings, and anecdotes from her experiences during the war.

Douglas was already living an unconventional life at the beginning of the war. A 1912 graduate of Wellesley, she married a con artist thirty years her senior and divorced him a year later when her father uncovered details of his financial fraud and forgery. In 1915 she moved to the then-small backwater town of Miami, Florida, and became the society editor for her father's newspaper, the *News Record* (later the *Miami Herald*). Sent to cover the story of the first woman in Miami to become a Yeoman (Female) when the Navy first opened enlistment to women, Douglas ended up taking her place: "I arrived at the ship and the next thing I knew I was sticking up my hand, swearing to protect and defend the United States of America from all enemies whatsoever. I guess they talked me into it." The recruiters, she said, "came on very strong."[2]

The Navy and Douglas suited each other poorly. Douglas quickly grew bored of stenography, typing, and filing, and rebelled by showing up late for work. She infuriated her commanding officers in the Miami reserve headquarters when she rewrote their correspondence to improve grammar and style. She was proud to have served, especially given seagoing ancestors from Cornwall and Maine, but at the end of a year she requested and received a discharge.

Naval service inspired Douglas to volunteer overseas with the Civilian Relief department of the Red Cross. Assigned in Paris at the end of the war, she wrote publicity releases and worked with war refugees. Her experiences with refugees and, after the Armistice, in the war-ravaged Balkans, found their way into her fiction after the war.

Douglas' literary legacy includes four novels and several book-length works of nonfiction; more than a hundred short stories and articles: forty of which were published in the *Saturday Evening Post*; two short stories which won O. Henry Awards; poetry; one-act plays; an autobiography, published in her ninety-seventh year; and her masterpiece *The Everglades: River of Grass*, an environmental and social work of nonfiction that has been compared to Rachel Carson's *Silent Spring* in its influence on public policy and environmental protection.

Douglas sprinkles nuggets of writing wisdom throughout her autobiography. To place her work in the *Saturday Evening Post*, she "sat down and deliberately studied the kind of story I thought [the editor] liked…. a success story with a noble main character, tangled up with a little sex and a few cuss words thrown in." If she got stuck on a story, she followed the advice of a neighbor to "Forget about the story. Go out. Get drunk, fall in love, do anything, but forget it for a while. Then come back and write it cold, all over again." Although she asserted that "you can't teach people to write if they can't write, and if they can write, they don't need teachers," her life belies the claim.[3]

Her journalist father taught her to check the facts in the news stories she reported. She found a New York agent who helped her place freelance work. She read voraciously and widely and filled her home in Coconut Grove with books of all kinds.

Douglas wrote through her nineties. At seventy-nine she became an environmental activist, created the nonprofit organization "Friends of the Everglades," and began lecturing and lobbying state and local legislatures. In 1993, President Bill Clinton awarded her the Presidential Medal of Freedom, citing her for "her crusade to preserve and restore the Everglades [which] has enhanced our Nation's respect for our precious environment, reminding all of us of nature's delicate balance."[4] Marjory Stoneman Douglas, Navy veteran of World War One, died in 1998 at the age of 108.

Jerri Bell is the managing editor for *O-Dark-Thirty,* the literary journal of the Veterans Writing Project. She retired from the Navy in 2008; her assignments included antisubmarine warfare in the Azores Islands, sea duty on USS *Mount Whitney* and HMS *Sheffield,* and attaché duty at the U.S. Embassy in Moscow, Russia. Her work has been published in a variety of journals and newspapers, including the *Washington Post,* and has twice been nominated for a Pushcart Prize. She and former Marine Tracy Crow are the co-authors of *It's My Country Too: Women's Military Stories from the American Revolution to Afghanistan.*

Notes

[1] Marjory Stoneman Douglas and John Rothchild, *Marjory Stoneman Douglas: Voice of the River* (Sarasota, FL: Pineapple Press, Inc., 2011), 185.

[2] Stoneman Douglas, *Voice,* 112-113.

[3] Stoneman Douglas, 169, 182, and 185.

[4] "About Marjory," Friends of the Everglades, May 21, 2020, https://www.everglades.org/about-marjory/

WWI Woman Marine Sergeant
Lela Leibrand

TRACY CROW, veteran

In 1977, the same year WWI Marine Corps veteran Lela Leibrand, perhaps better known as the mother of dancer and actress Ginger Rogers, was buried, I joined the Marines and began a career in public affairs, writing press releases for civilian media and articles for military newspapers and magazines, just as Marine Sergeant Lela Leibrand had in 1918.

Of course I didn't know this in 1977. I wish I had.

In 1977, I was toiling away on releases and articles under the sexist, watchful direction of a top enlisted man who walked around the press pit striking the side of his leg with a yardstick like a metronome to rush us young Marine reporters toward deadlines. For me, the only enlisted woman reporter in the pit, he meted out special praise such as, "You proved you're more than a good looking pair of legs, after all." In 1977, I was unaware, and no doubt so was he, that I was actually standing metaphorically and historically on the shoulders of Marine Sergeant Lela Leibrand.

In fact, I wouldn't discover any of this until nearly forty years later. For perspective it might help to remember that in 1977 women's military contributions weren't discussed during Parris Island boot camp classes on Marine Corps history. How to properly apply makeup, sure, but women's contributions during two major World Wars, Korea, and Vietnam? Omitted as if women had been nothing more than a footnote to history.

So my discovery of Marine Sergeant Lela Leibrand, among the first ten women to join the Marines in 1918 (referred to then as Marinettes), and who had been assigned to the Corps' Publicity Bureau, was nearly by accident. The discovery happened while researching for a new book, the brainchild of co-author Jerri Bell, a retired lieutenant commander. For sixteen months, we scoured through archives of diaries, oral histories, published and unpublished memoirs, depositions, and Congressional testimony for women's true stories told in their own words of their military experiences since the Revolutionary War.

As research led us to World War I...enter Marine Sergeant Lela Leibrand.

What we discovered by Leibrand was her article, "The Girl Marines," written in 1918 or 1919 for an unidentified publication. Her article breaks every rule of journalism, and I can only imagine the yardstick force of retribution that would have landed across my electric typewriter in 1977 had I dared to turn in such an energetic piece full of exclamation points and over-the-top personal bias. Yet with every read I fall more in love with Leibrand's unabashed exuberance and her eerie prescience.

Here's an excerpt from Leibrand's article that appeared in our forthcoming book, *It's My Country Too: Women's Military Stories from the American Revolution to Afghanistan* from the University of Nebraska Press/Potomac Books:

> Cherchez la femme! (Find the woman!) It is no longer a problem down at Headquarters in Washington. Girls, girls everywhere! And the Marines might just as well accustom themselves to us for we've come down among them to stay four years!

The moment your Marine Corps sent out the call for girls we flocked to the recruiting stations in every village, hamlet and town, eager to be one of that splendid body of men who have rendered such excellent account of themselves 'over there.'

. . . We are Marines! That says it all! . . . A sort of forewarning . . . do you gather a bit of our importance among you? We are not a fad by any means . . .

All this is just to let you know we are here, also to warn you that, whatever you do, remember the eagle eyes of the Marinettes are right upon you.

Watch your step![1]

Lela Leibrand (Rogers) testified before the House Un-American Activities Committee in 1947. The video is available on YouTube.

Tracy Crow is a literary agent and president/CEO of MilSpeak Foundation, a 501(c)3 nonprofit dedicated toward supporting and publishing the creative works of military veterans and family members through its imprints, MilSpeak Books and Family of Light Books. She is also the author/editor of six books to include the novella, *Cooper's Hawk: The Remembering;* the popular history, *It's My Country Too: Women's Military Stories from the American Revolution to Afghanistan* with co-author Jerri Bell; the award-winning memoir, *Eyes Right: Confessions from a Woman Marine;* and the breakthrough writing text, *On Point: A Guide to Writing the Military Story.*

Notes

[1] Jerri Bell, Tracy Crow, and Kayla Williams, *It's My Country Too: Women's Military Stories from the American Revolution to Afghanistan* (Lincoln, NB: Potomac Books, an imprint of the University of Nebraska Press, 2017), 82-84.

Equal Pay, Equal Benefits: Loretta Perfectus Walsh, the First Enlisted Woman in the U.S. Military

KAYLA WILLIAMS, veteran

When I enlisted in the Army in 2000, my pay was identical to that of all other servicemembers with my same rank and time in service. It did not occur to me at the time that this not only put the military in stark contrast to many other career fields but was also the result of trailblazing women who paved the way for the rest of us.

Loretta Perfectus Walsh was the first. Not the first woman to fight on behalf of America—that honor dates all the way back to the Revolutionary War—but the first to officially enlist, as a woman, earning equal pay and benefits. Earlier, women served either alongside their husbands or disguised as men in an era of, shall we say, more relaxed entrance examinations! Later, they could serve in one of the Nurse Corps—but without equal pay, benefits, and rank.

When the enormous need for personnel during WWI became clear, Navy Secretary Josephus Daniels took advantage of a loophole in the Naval Reserve Act to recruit women to fill clerical positions. The first woman to join was Loretta P. Walsh, a native of Olyphant, Pennsylvania, with secretarial skills. According to a 1917 *The New York Times* article, Walsh was sworn in as a chief yeoman after undergoing "the same rigid examination that is given to male applicants" and being assessed as "a fine specimen of womanhood."[1] The Navy was so unprepared for women to serve that she had to have her own uniform made, though she was issued a sabre and pistol.

Sadly, she died of tuberculosis before turning thirty, possibly because of contracting the Spanish flu during her time in the military. Her tombstone recognizes the import of her service, as the first "woman and patriot" enrolled in the Navy. She would not be the last: over eleven thousand women followed in her footsteps during WWI alone. Since then, over two million women have served in the U.S. military, gradually serving in a growing array of roles. More recently, the final remaining barriers were broken, and women can now serve even in combat arms positions—the first woman became an infantryman in 2017.

Walsh's historic enlistment took place before women were even allowed to vote—and yet she earned equal pay. It therefore seems fitting that the anniversary of her enlistment falls during Women's History Month: as we commemorate the histories of both WWI and women, let us honor her and her sisters in arms for all they helped make possible.

Kayla Williams was enlisted for five years and authored the memoirs *Love My Rifle More Than You: Young and Female in the U.S. Army* and *Plenty of Time*

When We Get Home: Love and Recovery in the Aftermath of War. Ms. Williams has a BA from Bowling Green State University and an MA from American University.

Notes

[1] "Philadelphia Woman Enlists in the Navy; Miss Loretta Walsh, Chief Yeoman, Said to Have Set Precedent for the World," *The New York Times* (The New York Times), accessed January 15, 2022, https://timesmachine.nytimes.com/timesmachine/1917/03/22/102323577.html?pageNumber=2.

Ellen N. La Motte's The Backwash of War.
Did a Censored Female Writer Inspire
Hemingway's Famous Style?*

CYNTHIA WACHTELL, PhD

Nearly everyone has heard of Ernest Hemingway. But you'd be hard-pressed to find someone who knows of Ellen N. La Motte.

People should know La Motte, the extraordinary WWI nurse who wrote like Hemingway before Hemingway. She was arguably the originator of his famous style—the first to write about WWI using spare, understated, declarative prose.

Long before Hemingway published *A Farewell to Arms* in 1929—long before he graduated high school and left home to volunteer as an ambulance driver in Italy—La Motte wrote a collection of interrelated stories titled *The Backwash of War.*

Published in the fall of 1916, as the war advanced into its third year, the book is based upon La Motte's experience working at a French field hospital on the Western Front.

"There are many people to write you of the noble side, the heroic side, the exalted side of war," she wrote. "I must write you of what I have seen, the other side, the backwash."[1]

The Backwash of War was immediately banned in England and France for its criticism of the ongoing war. Two years and multiple printings later—after being hailed as "immortal" and America's greatest work of war writing—it was deemed damaging to morale and also censored in wartime America.

For nearly a century, it languished in obscurity. But now, an expanded version of this lost classic that I've edited has just been published. Featuring the first biography of La Motte, it will hopefully give La Motte the attention she deserves.

HORRORS, NOT HEROES

In its time, *The Backwash of War* was, simply put, incendiary.

As one admiring reader explained in July 1918, "There is a corner of my book-shelves which I call my 'T N T' library. Here are all the literary high explosives I can lay my hands on. So far there are only five of them."[2] *The Backwash of War* was the only one by a woman and by an American.

In most of the era's wartime works, men willingly fought and died for their cause. The characters were brave, the combat romanticized.

Not so in La Motte's stories. Rather than focus on WWI's heroes, she emphasized its horrors. And the wounded soldiers and civilians she presents in *The Backwash of War* are fearful of death and fretful in life.

Filling the beds of the field hospital, they are at once grotesque and pathetic. A soldier slowly dies from gas gangrene. Another suffers from syphilis, while one patient sobs and sobs because he doesn't want to die. A ten-year-old Belgian boy is fatally shot through the abdomen by a fragment of German artillery shell and bawls for his mother.

War, to La Motte, is repugnant, repulsive, and nonsensical.

The volume's first story immediately sets the tone: "When he could stand it no longer," it begins, "he fired a revolver up through the roof of his mouth, but he made a mess of it." The soldier is transported, "cursing and screaming," to the field hospital.[3] There, through surgery, his life is saved but only so that he can later be court-martialed for his suicide attempt and killed by a firing squad.

After *The Backwash of War* was published, readers quickly recognized that La Motte had invented a bold new way of writing about war and its horrors. *The New York Times* reported that her stories were "told in sharp, quick sentences" that bore no resemblance to conventional "literary style" and delivered a "stern, strong preachment against war."[4]

The *Detroit Journal* noted she was the first to draw "the real portrait of the ravaging beast." And the *Los Angeles Times* gushed, "Nothing like [it] has been written: it is the first realistic glimpse behind the battle lines... Miss La Motte has described war—not merely war in France—but war itself."[5]

LA MOTTE AND GERTRUDE STEIN

Together with the famous avant-garde writer Gertrude Stein, La Motte seems to have influenced what we now think of as Hemingway's signature style—his spare, "masculine" prose.

La Motte and Stein—both middle-aged American women, writers, and lesbians—were already friends at the start of the war. Their friendship deepened during the first winter of the conflict when they were both living in Paris.

Even though they each had a romantic partner, Stein seems to have fallen for La Motte. She even wrote a "little novelette" in early 1915 about La Motte, titled *How Could They Marry Her?* It repeatedly mentions La Motte's plan to be a war nurse, possibly in Serbia, and includes revealing lines such as "Seeing her makes passion plain."[6]

Without a doubt Stein read her beloved friend's book; in fact, her personal copy of *The Backwash of War* is archived at Yale University.

HEMINGWAY WRITES WAR

Ernest Hemingway wouldn't meet Stein until after the war. But he, like La Motte, found a way to make it to the front lines.

In 1918, Hemingway volunteered as an ambulance driver and shortly before his 19th birthday, was seriously injured by a mortar explosion. He spent five days in a field hospital and then many months in a Red Cross hospital, where he fell in love with an American nurse.

After the war, Hemingway worked as a journalist in Canada and America. Then, determined to become a serious writer, he moved to Paris in late 1921.

In the early 1920s, Gertrude Stein's literary salon attracted many of the emerging post-war writers, whom she famously labeled the "Lost Generation."

Among those who most eagerly sought Stein's advice was Hemingway, whose style she significantly influenced.

"Gertrude Stein was always right," Hemingway once told a friend.[7] She served as his mentor and became godmother to his son.

Much of Hemingway's early writing focused on the recent war.

"Cut out words. Cut everything out," Stein counseled him, "except what you saw, what happened."[8]

Very likely, Stein showed Hemingway her copy of *The Backwash of War* as an example of admirable war writing. At the very least, she passed along what she had learned from reading La Motte's work.

Whatever the case, the similarity between La Motte's and Hemingway's styles is evident.

Consider the following passage from the story "Alone," in which La Motte strings together declarative sentences, neutral in tone, and lets the underlying horror speak for itself:

> They could not operate on Rochard and amputate his leg, as they wanted to do. The infection was so high, into the hip, it could not be done. Moreover, Rochard had a fractured skull as well. Another piece of shell had pierced his ear, and broken into his brain, and lodged there. Either wound would have been fatal, but it was the gas gangrene in his torn-out thigh that would kill him first. The wound stank. It was foul.[9]

Now consider these opening lines from a chapter of Hemingway's 1925 collection *In Our Time:*

> Nick sat against the wall of the church where they had dragged him to be clear of machine gun fire in the street. Both legs stuck out awkwardly. He had been hit in the spine. His face was sweaty and dirty. The sun shone on his face. The day was very hot. Rinaldi, big backed, his equipment sprawling, lay face downward against the wall. Nick looked straight ahead brilliantly . . . Two Austrians dead lay in the rubble in the shade of the house. Up the street were other dead.[10]

Hemingway's declarative sentences and emotionally uninflected style strikingly resemble La Motte's.

So why did Hemingway receive all of the accolades, culminating in a Nobel Prize in 1954 for the "influence that he has exerted on contemporary style," while La Motte was lost to literary oblivion?[11]

Was it the lasting impact of wartime censorship? Was it the prevalent sexism of the post-war era, which viewed war writing as the purview of men?

Whether due to censorship, sexism, or a toxic combination of the two, La Motte was silenced and forgotten. It's time to return *The Backwash of War* to its proper perch as a seminal example of war writing.

*A version of this article was first published in *The Conversation* on April 2, 2019. Reprinted with permission by Cynthia Wachtell.

Cynthia Wachtell is a research professor of American Studies and director of the S. Daniel Abraham Honors Program at Yeshiva University. She earned her PhD in the History of American Civilization at Harvard University. She is editor of *The Backwash of War: An Extraordinary American Nurse in World War I*, for which she wrote the first biography of Ellen N. La Motte, and author of *War No More: The Antiwar Impulse in American Literature from 1861-1914.*

Notes

[1] Ellen N. La Motte, *The Backwash of War: An Extraordinary American Nurse in World War I*, ed. by Cynthia Wachtell (Baltimore, MD: Johns Hopkins University Press, 2019), 135.

[2] La Motte, *The Backwash*, 9.

[3] La Motte, 96.

[4] "Backwash of War," *New York Times Book Review*, Oct 15, 1916, 432.

[5] La Motte, *The Backwash*, 6.

[6] La Motte, 64.

[7] Lyle Larsen, *Stein and Hemingway: The Story of a Turbulent* Friendship (Jefferson, NC: McFarland & Co., 2011), 32.

[8] Larsen, *Stein*, 28.

[9] Larsen, 115.

[10] Ernest Hemingway, *In Our Times* (New York: Boni & Liveright, 1925), 81.

[11] "The Nobel Prize in Literature 1954," NobelPrize.org, accessed January 15, 2022, https://www.nobelprize.org/prizes/literature/1954/hemingway/facts.

Josephine: Government Girl, 1918[1]

MARGARET THOMAS BUCHHOLZ

Josephine Lehman Thomas. She writes in her diary:

> June 1918: The gowns were of the real evening variety. I have gotten entirely over being shocked by such a display of arms and necks. After I have some new glad rags I am going to have some large pictures taken to astonish the Ionians by the general absentness of the upper part of my gown.[2]

My mother, Josephine Lehman Thomas, who spent the last half of her life in Harvey Cedars, New Jersey, was a "government girl" in Washington DC during WWI. She lived as a modern young woman who relished the wartime excitement of the nation's capital—with its soldiers, movie stars, and international celebrities. She kept a detailed diary from 1917 until the late twenties, when she was a researcher and ghostwriter for Lowell Thomas. Her entries offer a lively record of city life at the time. The following is excerpted from her story which appeared in *Washington History* in 1999 and was later expanded into my book, *Josephine: from Washington Working Girl to Fisherman's Wife.*

Josephine grew up on a farm in Michigan, graduated high school in 1915, and was hired by the *Ionia Sentinel*, the county's largest daily, and quickly worked her way up to reporter. The European war had started, and Jo was proofreading stories about U-boat sinkings, atrocities in Belgium, and the terrific loss of life in stalemate trench warfare. After the U.S. entered the war in April 1917, A *Sentinel* editorial exhorted its readers to "Wake up to the magnitude of the task before you. Wake up to the power and the far-flung resourcefulness of the enemy. Wake up to the fact that it is a man's fight and it is only just beginning."[3]

Soldiers off to War, March 24. Josephine writes:

> I went to a dance at Camp Laurel last night, a big military camp between here and Baltimore. There were six military five-ton trucks filled with girls from Washington, twenty-five going from our house. The six trucks went right down Pennsylvania Avenue and past the Capitol. It is a common sight to see truckloads of soldiers, but truckloads of girls are rather unusual. It was a farewell party for the boys who are leaving for France this week.[4]

But it was also a woman's fight.

President Wilson called on women to fight on the home front, and soon national women's magazines ran the government's full-page advertisements calling for "stenographers and typewriters" in the nation's capital.

Nineteen-year-old Jo, adventurous and idealistic, was caught up in the fervent patriotism as she read war dispatches and reams of government propaganda releases. She decided to take the civil service test and recorded in the handwritten diary she'd begun in May: "Am informed I will come through with flying colors [...]Have started to look up railroad schedules and places of interest in Washington." Six weeks later, a telegram from the War Department confirmed her appointment as a clerk in the Office of the Chief of Ordnance.

On February 19, 1918, Jo arrived at ten-year-old Union Station and entered the waiting room to search for the war housing information booth. As they did for thousands of unprepared newcomers, the housing coordinators helped her find a room for the night two blocks from the station.

A few days later she sent *The Sentinel* a detailed report:

At the station, I obtained the address of a rooming house only two blocks from the depot but I didn't want to attempt getting lost so let a taxicab driver soak me for 40 cents. The house was overflowing, but the lady of the house was kind enough to rout out a lieutenant who stays there and make him carry down a small bed so she could bunk me in the parlor, which I considered extremely kind of her. Thanked the Lord I am gifted with a strong constitution and so managed to get up for a formal breakfast and report to work by nine o'clock.

I was sworn into the Civil Service and sent over to the ordnance war building for work. A thin, middle-aged man and a big, young man argued over which should get me, both wanting a stenographer with business experience. I watched the fray in silence, mentally rooting for the young man. He won. My job is in the supply division of the ordnance department. The building is a mammoth new one and I have to have a pass to get in and out. This is the hardest department to work in as they are so busy. The girls don't get many holidays but pay and promotions are good.[5]

The following day, she used the room registration office at the War Department to find lodging at 1415 Massachusetts Avenue, N.W.:

I can't tell you how much I like it.[...]We have 23 girls living here and either two, three or four to a room, so don't worry about my being lonely. The rooms are large, so we don't mind it. I have two others in my room, Miss Dell

Brokaw from Illinois, a tall, stunning blond, and Miss Grace Leonard from New York, also a big girl, with the most beautiful gown you ever saw. They call us the "Big Four Minus One," "Amazons" and other endearing names. [My mother was five feet, nine inches tall.] All the girls are the finest kind, splendid, and very congenial. We take our breakfasts and dinners here and lunch downtown....

And soldiers! There are about fifteen camps within a short radius of Washington. It seems the soldiers I have seen would make an army big enough to demolish the Kaiser in a day. Sailors, Marines, aviators, cavalry, infantry and artillerymen, ordnancers, plain desk-holder-downs and many IWTGBCs. The initials stand for a certain order who are going to petition to wear buttons that say 'I Want To Go But Can't.'

My hours are from 9 to 4:30 with double pay for overtime. There is a movement to lengthen the federal employees working day to eight or nine hours and all the department clerks went to protest, but even if they do give us longer hours I can realize that we still are not sacrificing much.[6]

Jo joined the surge of new government employees that swelled Washington's wartime population from approximately 350,000 in April 1917 before the war to approximately 526,000 a year later. A hotel room was an unheard-of luxury. A single room in a boarding house or private home rented for nearly as much as an entire house had before the war. Three or four people occupied a room intended for one, and many young women had to share a bed.

From the Capitol Steps on April 6, she writes:

Today the third Liberty Loan campaign was launched [...]. At four o'clock, although we were not supposed to have left, we took advantage of Mr. Brady's absence and walked out en masse [...]. By dint of much pushing and shoving, I jostled my way to Pennsylvania Avenue through the dense crowd. My height was in my favor in looking over the crowds, and I held a place near the edge of the throng. All affairs of war were pushed from my mind because I was soon to see probably the best-known people in the country. [...] Mary Pickford appeared in the first carriage. She is very tiny [...]. She looked very sweet and charming. [...] The Charlie [Chapin] of the movies, minus diminutive derby, enormous trousers and shoes and the ever-present mustache appeared as an attractive, dark-haired young man with an unusually likable smile, unhampered by the mustache. [...] Douglas Fairbanks, in the same car, came up to expectations by looking exactly like I always supposed he did. His famous grin was all wool and a yard wide. He was tanned and jocular and slangy and a hit with the crowd. [...] Overhead were the best aviators dipping, curving, whirling and showering thousands of Liberty Bells upon the crowd. It was a fitting beginning for the Liberty

Loan drive and a celebration that I shall never forget. I had an hour's worth of work to catch up when I returned to the office. [7]

By early June, Washington looked like a military base, and military activities permeated almost all aspects of life. Even though President Wilson still took occasional walks to his bank or a shop, the two White House gates were guarded by military policemen with rifles bayonet-ready; no longer could casual sightseers walk up to the Executive Mansion. Soldiers bunked in tents along the riverbank, keeping watch over the Potomac rail bridge. Further down the river, the polo field was converted to a landing strip. The section of Potomac Park not covered by temporary government buildings sprouted vegetables. Construction of the Lincoln Memorial continued, and plans for "speedily erected, readily taken down" workers' housing between it and the river were on the drawing board. Jo recorded in her diary:[8]

> Mrs. Dudley had a dance at the house last night. Seventy-five here and excellent time. Sat out a dance with Sergeant Kenyon of the British Royal Flying Corps who is in Washington as an instructor. As soon as his machine arrives he is going to do some stunts over the city. Nearly every day I see daring exhibition aeroplane flights over the city and at night the sky is shot through with many searchlights playing on the dome of the Capitol. Some thrilling thing is going on all the time and the city is full of soldiers, sailors, aviators, etc. [9]

The War Department's Committee on Training Camp Activities took enough of an interest in its workers' leisure-time activities to issue a warning in the August 14, 1918, edition of the *Evening Star* against "pick-up soldier acquaintances." The warning was not intended to reflect on the character of the servicemen, who, it stated, were "clean and upstanding."[10] Rather, the War Department did not want its girls conversing familiarly with men in uniform unless introduced by a mutual friend or vouched for by a community organization.

A new vocabulary was heard. Civilians became accustomed to military talk and used words such as "liaison, communique, propaganda, and sector" in their conversations; phrases such as "over the top" and "carry on" entered common usage. Uniformed men mingled with smartly dressed crowds in restaurants, theaters, and dances. Weary-eyed English, French, and Italian officers, many lame or battle-scarred, were billeted in private homes. Battalions of soldiers were being trained, and more and more Americans in khaki or Navy blue flowed through the camps surrounding Washington on their way "over there."

Walter Reed April 8, 1919:

I have been filling the time in the usual methods, and have added a new role, that of entertainer to the Walter Reed Hospital boys [...]. Across the table from me was a boy who had been gassed severely. He hadn't been able to get his spectacles off in time to put on his gas mask, he told us.[11]

Citizens stood in line everywhere. Lines formed to read war bulletins in front of the *Evening Star* building at 1101 Pennsylvania Avenue, N.W. People queued up for restaurants and street-corner food carts, at banks and in the imposing green marble lobby of the new post office next to Union Station, at theater and movie box offices, at butcher shops and greengrocers. Lines even formed around public telephone exchanges as frustrated residents besieged overworked operators. Hundreds of private automobiles brought to the city by the dollar-a-year men (industrialists donating their services to the government during the emergency) challenged the trolleys and remaining horse-drawn carriages, congesting the downtown area.

Flags flew everywhere, fluttering high against the sky, draped over doors, or falling in folds from windows. Patriotic posters were at eye level on street corners and in the multitude of small grassy triangle parks. Beginning in August a flag-draped siren blew every day at noon, and all Washington stopped work to pray for victory.

In September, Spanish influenza reached eastern seaboard cities, after ravaging armies in Europe. The disease entered Washington from surrounding camps during the first week of October with a force that staggered federal office shifts. More than half the streetcar operators became sick. Store hours were limited; saloons, schools, churches, and theaters were closed as public gatherings were prohibited and social activities were strictly curtailed.

October 1918: The girls in Jo's house planned an elaborate Halloween party, but, as the *Evening Star* reported that year, Halloween would be "overshadowed by war and influenza." They didn't have their costume ball until after the Armistice. Jo wrote a letter to the Sentinel about the epidemic, "Out of the twenty-five girls at my boarding house, about twenty were ill, and it was a young hospital all by itself."[12]

Government employees were ordered out of buildings for fresh-air breaks in an elemental attempt to foil the infection.

District Commissioner Louis Brownlow recalled the epidemic's uncontrollable horror: "A girl had called to say that she and three other girls had a room together, that two of the girls were dead, another was dying and she was the only one not stricken; would I please get some help there."[13] The policeman he sent to the house discovered all four girls dead in their room. Approximately 35,000 Washingtonians caught the flu, and ten percent of them had died by the time the outbreak subsided in early November.

Margaret Thomas Buchholz is co-author of *Great Storms of the Jersey Shore*, author of *New Jersey Shipwrecks: 350 Years in the Graveyard of the Atlantic*, *Josephine: From Washington Working Girl to Fisherman's Wife*, and editor of *Shore Chronicles: Diaries and Travelers Tales from the Jersey Shore 1764–1955*, *Long Beach Island Chronicles*, *LBI Reader*, and *Island Album*. Her essays about the shore have also been included in anthologies and collections. Buchholz was publisher of the LBI newspaper the *Beachcomber* from 1955 to 1987. She lives in her childhood home in Harvey Cedars, New Jersey, on Barnegat Bay, where her family has been coming since the Civil War.

Notes

[1] This essay contains excerpts from Josephine Lehman Thomas and Margaret Thomas Buchholz, "Josephine: The Washington Diary of a War Worker, 1918-1919," *Washington History*, 1998-1999, pp. 4-23. Reprinted with permission by Margaret Thomas Buchholz.

[2] Quoted in Josephine Lehman Thomas and Margaret Thomas Buchholz, "Josephine: The Washington Diary of a War Worker, 1918-1919," *Washington History*: 1998-1999, 14.

[3] Quoted in Thomas and Buchholz, "Josephine," 5.

[4] Josephine Lehman Thomas. *The Diary of Josephine Lehman Thomas*. Private collection belonging to Margaret Thomas Buchholz.

[5] Quoted in Thomas and Buchholz, "Josephine," 11.

[6] Quoted in Thomas and Buchholz, 13.

[7] Quoted in Thomas and Buchholz, 15.

[8] Quoted in Thomas and Buchholz, 8.

[9] Quoted in Thomas and Buchholz, 14.

[10] Quoted in Thomas and Buchholz, 14.

[11] Quoted in Thomas and Buchholz, 22.

[12] Quoted in Thomas and Buchholz, 9.

[13] Quoted in Thomas and Buchholz, 9.

Aline Kilmer:
When the War Poet's Wife is a Poet Too

PETER MOLIN, PhD, veteran

Today, poets such as Jehanne Dubrow, Elyse Fenton, Amalie Flynn, and Lisa Stice, as well as fiction authors like Siobhan Fallon and Andria Williams, all spouses of military men, portray the contortions on domestic life and feminine sensibility wrought by war. The possibilities for pride and happiness are real but tenuous, infused everywhere by the realities of separation, divided loyalties, fear, and mortality. These authors' words convey the urgency and nuance of a war wife's uncertainty as she finds tranquility and self-worth vexingly dependent on her husband.

Writing openly about self and marriage requires enormous courage. Perhaps it was ever so, or even more during the years of WWI. Aline Murray Kilmer was the wife of American poet Joyce Kilmer, most famous for his poem "Trees." Joyce's status as a WWI poet is secured by poems such as "Rouge Bouquet" and "The Peacemaker," both written shortly before his death in 1918 at the Second Battle of the Marne. The fact that Aline was also a poet is not well-known, and the possibility that her work dealt with WWI is less considered. Was Aline Kilmer a war poet? What are her war poems like? What connects her to today's military spouses?

Aline, born in 1888 in Norfolk, Virginia, first met Joyce in high school in New Brunswick, New Jersey. Aline and Joyce married in 1908, and together they had five children. Rose, one of their daughters, died just before Joyce deployed to France. Aline published three volumes of poetry and an essay collection by 1925, plus two volumes of children's poetry and an edition of selected poems before the end of the decade. Many of her poems and essays were first published in prominent magazines of the day such as, among others, *Harper's*, *Good Housekeeping*, and *The Smart Set*. But after 1929, nothing appeared before Aline died in 1941. Today, none of Aline's books are in print, and though many of her poems are on the Internet, they have not attracted scholarly appraisal or popular acclaim.

Exploration of Aline's life—particularly her relationship with Joyce—and her poetry—especially its reference to war—is difficult, for few of her poems refer explicitly to Joyce or war. Aline never wrote a memoir, nor did she publish reminiscences of her late husband—famous for "Trees" and a nationally known literary critic— before joining the Army in 1917. For reasons unknown, she destroyed the correspondence with Joyce while he was in France. In her poetry, Aline often mentions her children by name, but never Joyce's name, though a vaguely defined and occasionally recurring "you" suggests that at least some of them were aimed at her husband. The poems aren't biographies, but it's still hard to read them and not speculate how they helped Aline manage the fear, doubt, grief, and confusion occasioned by war and her husband's death.

Aline's first published volume of poetry, *Candles That Burn*, appeared in 1919. Many of its poems, all short lyrics expressed in conventional verse forms, had previously appeared in magazines, some predating Joyce's military enlistment. Published within two years of the deaths of Joyce and daughter Rose, *Candles That Burn* contains references to Rose but few to Joyce. The poem, "I Shall Not Be Afraid," provides one striking exception; Aline may have written it after Joyce deployed, but before he died:

> I shall not be afraid any more,
>   Either by night or day;
> What would it profit me to be afraid
>   With you away?
> Now I am brave. In the dark night alone
>   All through the house I go,
> Locking the doors and making windows fast
>   When sharp winds blow.
> For there is only sorrow in my heart;

> There is no room for fear.
> But how I wish I were afraid again,
>   My dear, my dear![1]

"In Spring" may be addressed to either Joyce or Rose or both, and it may, like "I Shall Not Be Afraid," be about physical distance, rather than separation by death:

> I do not know which is worse when you are away:
>   Long grey days with the lisping sound of the rain
> And then when the lilac dusk is beginning to fall
>   The thought that perhaps you may never come back
>   again;
> Or days when the world is a shimmer of blue and gold,
>   Sparkling newly all in the dear spring weather,
> When with a heart that is torn apart by pain
>   I walk alone in ways that we went together.[2]

Only in "Christmas" are Aline's two recent losses specifically mentioned:

> I shall sit alone by the fire and see
> Ghosts of you both come close to me.
> For the dead and the absent always stay
> With the one they love on Christmas Day.[3]

One of the most interesting poem from *Candles That Burn* is "To a Young Aviator:"

> When you go up to die
>   Some not far distant day,
> I wonder will you try
>   To tear your mask away,
>   And look life in the eyes
>   For once without disguise?
> Behind your mask may hide
>   What treacherous, covered fires!
> What hidden, torturing pride!
>   What sorrows, what desires!
>   Whatever there may be
>   There will be none to see.
> Yet I think when you meet
>   Death coming through the skies,
> Calmly his face you'll greet,
>   Coldly, without surprise;
>   Then die without a moan,
>   Still masked although alone.[4]

The poem suggests the aviator's bravado is a mask covering not cowardice—for the aviator is clearly brave—but loneliness. It also suggests that loneliness is not as important as the fact that the aviator dies without letting his true self be known. Joyce Kilmer served on the ground as an intelligence analyst in an infantry regiment and not in the air, so he cannot literally be the "young aviator" whom Aline addresses, but her description of the airman's death possibly alludes to Joyce, who had pinned to fight for his country on the frontlines instead of working behind the front lines. scenes. Was this pilot like the Joyce that Joyce wanted to be but feared he wasn't? Is Aline saying that her husband died "still masked although alone?"

Even before Joyce joined the Army, Aline might have resented the prescribed social roles that kept her at home while Joyce traveled the country giving lectures on poetry after the success of "Trees." It's not impossible to imagine that Aline was unhappy with Joyce's decision to join the Army at age 30 at a lower salary than he was making as an author and speaker, not to mention leaving her at home with five children. When Joyce was asked how his wife felt about his decision to leave his career and family to fight in France, he reportedly replied "She's game."[5] In light of Rose's death and Joyce's imminent one, Aline's feelings must have been more complex.

It's also hard to ignore another biographical source of tension: Aline and her mother-in-law, Annie Kilmer, who wrote three memoirs about her son, appear to have not gotten along. In her memoirs, Annie Kilmer's references to Aline are infrequent and terse. In turn, Aline's poems and essays never mention Annie.

Though addressed indirectly, tense currents animate the poems in 1921's *Vigils*. In "Light Lover," for example, Aline appears to accuse Joyce of abandoning her: "Take your restless heart to the restless sea,/ Your light, light love to a lighter lass."[6]

In "If I Had Loved You More," however, Aline indicts herself:

If I had loved you more God would have had pity;
  He would never have left me here in this desolate
    place . . .
For you were always the lover and I the friend.[7]

Joyce, in a letter, echoed nearly the same sentiment, that his ardor for Aline outpaced hers for him. Though expressed light-heartedly, perhaps it was true. Or was it perhaps Aline who did something awful to ruin the marriage? She appears to say so in the ominously-titled "Shards:" "I can never remake the thing I have destroyed . . . I did a terrible thing."[8]

Anguish and guilt pervade "Atonement," another *Vigils* poem:

I wake in fear and put out my hand to find you
  With your name on my lips.
This should atone for the hours I forget you.
  Take then my offering, clean and sharp and sweet,
An agony brighter than years of dull remembrance.
  I lay it at your feet.[9]

Whether it was Rose's death or Joyce's death that pained Aline more is unclear. What is clear is that Aline's sense of herself as marked by evil is profound. She also reiterates a belief that her sins, crimes, or even simple knowledge of life's dark side is relational, caused by and manifested in mutual misunderstanding with someone close. "Perversity" states the sentiment is stated most directly:

All my life I have loved where I was not loved,
And always those whom I did not love loved me.
Only the God who made my wild heart knows
Why this should be.
Oh, I am strange, inscrutable, and proud;
You cannot prove me though you try and try
I'll keep your love alive and wondering
Until you die.[10]

It also appears in "Bound:"

If I had loved you, soon, ah, soon I had lost you.
  Had I been kind you had kissed me and gone your
    faithless way.
  The kiss that I would not give is the kiss that
    your lips are holding:
Now you are mine forever, because of all I have
    cost you.
You think that you are free and have given over your
    sighing,
  You think that from my coldness your love has
    flown away:
  But mine are the hands you shall dream that your
    own are holding,

And mine is the face you shall look for when you are dying.[11]

The lyrics of "Knowledge" depict stunning images of Aline's peculiar sense of her damnation:

> Some learn it in their youth,
> Some after bitter years:
> There is no escape from the truth
> Though we drown in our tears.
> Many die when they see
> That the terrible thing is true
> But it was always easy for me:
> I always knew.[12]

Aline's 1925 *The Poor King's Daughter and Other Verses* doesn't feature the intense self-laceration of the best poems in *Vigils*. Instead, its poems suggest that Aline had become more resigned, her fascination with herself as stained by darkness more subdued, and her contemplation of mortality formerly centered on others now directed at her own in "You Asked Me Not to Die:"

> . . . Nay, Sweet, to-morrow
> Your flower-like beauty may have failed and fled,
> And I shall weep you dead;
> Then rise to face the grim and hooded years,
> Each with his vase of tears,
> That moves majestically by;
> Till the little I had of beauty will be but a withered mask,
> And the little I had of wit will be bitter and dry.
> Dear, you do not know what it is that you ask.
> How can you love me and bid me not to die?[13]

References to Joyce, Rose, and her other children fade, and in "Tournament" Aline portrays herself locked in a titanic-but-doomed battle with the elements:

> I have fought with stars in their courses
> and dreamed I have won,
> I have charged full tilt with my levelled lance
> straight into the flaming sun
> And because of the darkness that swallowed me I
> have dreamed that the fight was done.[14]

In 1923 came *Hunting a Hair Shirt, and Other Spiritual Adventures*, a volume of essays and articles that look, contra its forbidding title, light-heartedly at middle-class domestic and literary foibles without mentioning war or Joyce. After 1925, she published only two volumes of children's poetry, as yet unexamined, and a collection of selected poems. Therefore, a critic's judgement must primarily be made on *Candles That Burn, Vigils,* and *The Poor King's Daughter.*

It's not impossible to imagine that after her husband's death, Aline's emotions encompassed more anger than grief, or that they combined grief with feelings of being cheated and abandoned. Aline was religious, and her poems often express spiritual longing. Both Aline and Joyce converted to Catholicism before the war, which helps explain the Manichean cosmology that runs through Aline's poems. She saw her soul, and perhaps her behavior, as the site of conflict between good and evil, or, perhaps, between the emotions she felt and those she felt comfortable expressing.

The best poems are not shaped by a heaven-or-hell binary, however. Instead, they are inwardly rooted in a complex and ambiguous psychology, grounded in a perverse image of herself as different, haunted, and doomed. They may not entirely escape self-pity or other forms of melodramatic representation, but their striking words and images ensure that once read, they are not soon forgotten.

How to understand Aline Kilmer as a war poet herself? Not easily and not clearly, but not impossibly, either. Aline's vague and ambivalent allusions to her husband Joyce complicate the traditional notion of the dutiful soldier's wife who dedicates all thoughts to her husband's valor even after he has died. Had our contemporary military spouse poets lived in a different time of war, say some one hundred years ago, and been forced to contemplate their husbands' deaths, who knows that they wouldn't have used the tactics of oblique reference and suggestiveness that generates the impressive power of Aline Kilmer.

Peter Molin teaches in the English Department Writing Program at Rutgers University, New Brunswick, New Jersey. A retired Army officer, he served in Afghanistan in 2008-2009 as an advisor to the Afghan army. He has written extensively on 19th-century American literature and 20th and 21st century war-writing. His blog, Time Now, has thousands of followers.

Notes

[1] Aline Kilmer, "I Shall Not Be Afraid" in *Candles That Burn* (New York: George H. Doran Company, 1919), 34.

[2] Kilmer, "In Spring," in *Candles That Burn*, 35.

[3] Kilmer, "Christmas," in *Candles That Burn*, 37.

[4] Kilmer, "To a Young Aviator," in *Candles That Burn*, 49.

[5] John E. Covell, *Joyce Kilmer: A Literary Biography* (Brunswick, GA: Write-Fit Communications), 136.

[6] Aline Kilmer, "Light Lover," in *Vigils* (New York: George H. Doran Company), 34.

[7] Kilmer, "If I Had Loved You More," in *Vigils*, 15.

[8] Kilmer, "Shards," in *Vigils*, 32.

[9] Kilmer, "Atonement," in *Vigils*, 16.

[10] Kilmer, "Perversity," in *Vigils*, 31.

[11] Kilmer, "Bound," in *Vigils*, 44.

[12] Kilmer, "Knowledge," in *Vigils*, 45.

[13] Aline Kilmer, "You Asked Me Not to Die," in *The Poor King's Daughter and other Poems* (New York: George H. Doran, Company), 39.

[14] Kilmer, "Tournament," in *The Poor King's Daughter and other Poems*, 45.

Letters That You Will Not Get:
Women's Voices from the Great War

SUSAN WERBE

December 1, 2018. Three years after I first envisioned a song cycle based on women's writings from both sides of the conflict and set to contemporary music, *Letters That You Will Not Get: Women's Voices from The Great War* had its premiere performance in New York at The Crypt, Church of the Intercession on Manhattan's Upper West Side. The libretto that I co-created with Kate Holland, who also directed the performance, was created from women's writings—American, British, Irish, French, and German. A British prime minister's daughter, an American poet, a Parisian factory worker, a German artist, among others. Women who set down their experiences in writing—both on the front lines and the home front. Daughters, mothers, wives, lovers who lost loved ones, experienced a shattered world, called upon both physical and emotional reserves previously unused.

WWI is considered to be the most literary of all wars. Until recently, writings from that war have been examined and studied as a male experience. However, my interest in creating performance pieces—a short dance work, set to commissioned contemporary music, *The Great War Theatre Project: Messengers of a Bitter Truth,* a multi-media theatre piece, and now music—has been to include women's writings to give each performance piece a balanced and nuanced lens through which to examine this history. Expressions of those womens' experiences were as forceful, powerful, and authentic as men's, even as their roles differed.

After the dance and theatre pieces ended, I found there remained a wealth of women's writings that had not been used—poetry, letters, journals, memoirs, diaries—responses to the war written during the conflict as well as reflections on their war experiences in the decades that followed. I felt strongly that a song cycle based solely on women's writings would bring these missing voices into the war's narrative. These voices include a range of responses from women of varied nationalities and backgrounds.

Upper-class English women who embraced the war, even as they sent their sons to the Western Front to fight and to die: "Violet passed on news from George: He said that up to date it had all been the most glorious fun."[1]

Snatches of letters from a French country woman: "My beloved Paul, Take care of yourself so that nothing will happen to you . . . "[2]

The anguish of a bereaved German mother:

Where are my children now?
What is left to their mother?
One boy to the right and one to the left,
My right son and my left son,
As they called themselves.
Where are my children now?
One dead and one so far away[3]

Poets of verse considered jingoistic and doggerel by British poet Jesse Pope:
Shining pins that dart and click

In the fireside's sheltered peace
Check the thoughts that cluster thick—
20 plain and then decrease.
He was brave—well, so was I—

Keen and merry, but his lip
Quivered when he said good-bye—
Purl the seam-stitch, purl and slip.
Never used to living rough,

Lots of things he'd got to learn;
Wonder if he's warm enough—
Knit 2, catch 2, knit 1, turn.
Wonder if he's fighting now,

What he's done an' where he's been;
He'll come out on top somehow—
Slip 1, knit 2, purl 14.[4]

This doggerel so infuriated Wilfred Owen, states Owen scholar Dr. Jane Potter in her paper presented at the Glasgow conference, entitled "Wait, Weep and Be Worthy?" paper presented at the Glasgow Conference that he felt:

> [...] compelled [...] to dedicate 'Dulce et Decorum Est' in his original manuscript 'To Jessie Pope, etc', then in a subsequent version 'To a Certain Poetess' before deleting the 'dedication' altogether in favour of the more cryptic reference to 'my friend' in the final lines that precede 'the Old Lie.'[5]

Women's Voices from the Great War, posits the idea that Pope's poetry today can be read in irony. And indeed that is how we used it in Letters' libretto.

Another acclaimed British war poet, Siegfried Sassoon, expressed in a poem his bitterness towards women who could not begin to understand what soldiers experience in the frontlines in "The Glory of Women":

> You love us when we're heroes, home on leave,
> Or wounded in a mentionable place.
> You worship decorations; you believe
> That chivalry redeems the war's disgrace.
> You make us shells. You listen with delight,
> By tales of dirt and danger fondly thrilled.
> You crown our distant ardours while we fight,
> And mourn our laurelled memories when we're killed.
> You can't believe that British troops "retire"
> When hell's last horror breaks them, and they run,
> Trampling the terrible corpses—blind with blood.

> O German mother dreaming by the fire,
> While you are knitting socks to send your son
> His face is trodden deeper in the mud.[6]

"The Glory of Women" worked wonderfully well in the theatre piece, *Messengers of a Bitter Truth*, as part of the narrative that formed the script, but *Letters* illustrates women's experiences. Women did "make shells" but were also blown up in industrial accidents in the munitions factories. German mothers did indeed "knit socks," but so many also mourned. Visual artist Käthe Kollwitz, whose younger son Peter was killed in the first months of the war, created powerful art that mourned the war dead, and represented through her art the women who suffered loss.

From the outset I felt strongly that these women's words had to be set to music by a woman composer. Rhode Island-based composer Kirsten Volness engaged deeply with the libretto and created extraordinary music that was both lyrical and discordant, both modern and reminiscent of the early 20th century, and finally elegiac as the singers performed the last song, adapted from the poem "There Will Come Soft Rains" by American poet Sara Teasdale:

> There will come soft rains and the smell of the ground,
> And swallows circling with their shimmering sound;
>
> And frogs in the pools singing at night,
> And wild plum trees in tremulous white,
>
> Robins will wear their feathery fire
> Whistling their whims on a low fence-wire;
>
> And not one will know of the war, not one
> Will care at last when it is done.
>
> Not one would mind, neither bird nor tree
> If mankind perished utterly;
>
> And Spring herself, when she woke at dawn,
> Would scarcely know that we were gone.[7]

The night of Letters' *premiere*, we gathered in the Crypt, the perfect venue—both acoustically and atmospherically. Three sopranos, a mezzo soprano, a cellist and a violinist (both women) brought the work beautifully to life. "We didn't even need to say it, but the fact that this piece was start to finish created by and performed by women, some of whom have been gone for many years, is just so important," reflected mezzo soprano Caitlin McKechney.

The audience and the performers were on a journey together, engaging with women who, 100 years ago, were profoundly affected by the 20th Century's first global conflict.

Susan Werbe is an independent scholar focused on the social and cultural history of early 20thC England and WWI and creator of several performance pieces based on original writings from the conflict. She was executive producer and dramaturg for *The Great War Theatre Project: Messengers of a Bitter Truth*, performed in Boston and New York (USA) and Letchworth (UK). She is co-librettist for the song cycle and chamber opera *Letters That You Will Not Get: Women's Voices from the Great War*. Susan has served as presenter or panelist at international conferences on WWI and on war and the creative arts.

Notes

[1] Adam Hochschild. *To End All Wars: A Story of Loyalty and Rebellion, 1914-1918* (New York, Houghton Mifflin Harcourt, 2011), 106.

[2] Martha Hanna. *Your Death Would Be Mine: Paul and Marie Pireaud in the Great War* (Cambridge, Massachusetts, 2006), 63.

[3] Käthe Kollwitz. "Diaries," trans. Henriëtte Kets de Vries. *Mother's Arms: Kathë Kollwitz's Women and War* (Northampton, MA: Smith College Museum of Art), 2016.

[4] Margaret R. Higonnet, ed. *Lines of Fire: Women Writers of World War 1* (New York, Plume, 1999), 477.

[5] Jane Potter, "Women, Poetry & War" (presentation Wait, Weep and Worthy? Women and the First World War. A Centenary Festival as Part of the Human Festival, The Glasgow Women's Library, Glasgow: November 2015).

[6] Jon Silkin, editor. *The Penguin Book of First World War Poetry*. Second edition. (Harmondsworth, Middlesex, England: Penguin Books 1981), 132.

[7] Sara Teasdale. *Flame and Shadow*. (New York, The Macmillan Company, 1920), 89.

part four

WWI MATTERED FOR BLACK LIVES

Iraq and Afghanistan Deployment
Inspired by
Ida B. Wells' WWI Fight

JASMINE WALKER MOTUPALLI, veteran

As a veteran of both the Afghanistan and Iraq wars, where I served as an intelligence officer, I often think about how different my life might have been had I lived in a different period of history. A quick look at most American wars tells me that I would have never been an officer. I would have never been on the front line. But I do know that I would have done whatever I could to fight the fight. As a black woman during WWI, perhaps I might have been supporting the war effort at home by working in factory jobs vacated by men. Perhaps I would have chosen to prove my patriotism by serving as a nurse for the segregated black units in the American Expeditionary Forces in Europe.

But, in the face of inequality, I tend to be a bit rebellious and, even in 1914-1918, I'm pretty sure I would have tried to break the ceiling even further. During this time, I would have followed one of my heroes, Ida B. Wells, journalist, suffragist, editor, and activist. She fought for voting rights and equal treatment for women who looked like me. During WWI, she sold Liberty Bonds and distributed care packages to black soldiers. But her fight didn't stop there. She worked to defend black men falsely accused of crimes throughout the country. Following the Houston Race Riot of 1917, she fiercely protested the hanging of thirteen black soldiers hanged by a military court without recourse to appeal or review by the president.

I have my father to thank for everything. He was both a feminist and a football star. He said that whatever era I was born into, I should never allow the color of my skin or my gender to dictate my path. He was the first to encourage my obstinate character. He said that I would become what I wanted, not despite, but thanks to the power of who I am—an African American, a woman, a human, a fighter, a leader, and a writer.

Here's an extract from my in-progress memoir about my time as an undergraduate at West Point, as a soldier in Iraq and Afghanistan, and my path to become a major and a West Point professor. During WWI, my current career and military contribution would have never been possible. All the same, African American women like Ida B. Wells made a difference and fought like soldiers. My debt is to women like her.

This excerpt comes from the chapter called "The Worst Day," the day a bounty was put on my head in Afghanistan as the only female Company Commander in my Brigade Combat Team and one of my soldiers had almost died after severe battle wounds. Everyone was down and I had to boost morale to keep them going. One of the lieutenants wanted us to pull out.

I promise, this is the only time that I've given a movie-like speech in my entire 30+ years of life. I also promise that I hated every minute of this speech. In retrospect, I feel like everyone should get one in his or her lifetime.

"Your platoon is looking to you right now for the lead on how to behave. You can't be doing this. WE don't fall apart." I pointed back and forth between my command and this lieutenant.

The lieutenant tried to speak again but I cut him off.

If you want to fall apart, you do it with me. You do it here and now. Then you take it like a leader, and you go in front of your platoon with me. All together . . . You think this is the last time something like this is going to happen? It's not . . . But you know what? We're STILL going to make sure our equipment is mission-ready and we're STILL going to do our missions. The Afghans are soldiers. WE are soldiers. We will all complain and moan—all of us. But once we're told to do it and drive on, that's what WE do. The same as I'm telling you to drive on right now. WE . . . DON'T . . . FALL . . . APART."

I have never hated myself more than I did at that moment. This lieutenant was a scared 22-year-old kid who had just had the worst day of his life. I wanted to hug him. I wanted to hug all of them. I wanted to cry and mourn and be a human being. What I wanted didn't matter. We had to get through a year of this and we wouldn't make it if we fell apart after every firefight. I would do what had to be done, say what had to be said, to get my company through this successfully. I would fight. I would fight against this enemy. I would fight for my Soldiers. I would fight for their lives and their souls, even at the expense of my own.

Over the next year, there would be hundreds more patrols within Pul-e-Alam District in Afghanistan, dozens more firefights with Taliban and Haqqani cells, some that I would be involved in and others that I wouldn't, and even a few more bounties placed on my head by Christmas. But the truth remained; nothing would leave me more shaken, than when one of my soldiers came in wounded. I realized very quickly that I didn't care about my own safety at all.

Jasmine Motupalli served for thirteen years as a U.S. Army officer, most recently as a data scientist at the Pentagon before transitioning to the tech industry. She also worked as an assistant professor at West Point, directing Engineering Statistics and Business Analytics courses. Her research included topics in STEM education, predictive analytics on veteran homelessness, and Unmanned Aircraft Systems. Before that, Jasmine served for eight years as a U.S. Army intelligence officer, leading teams in Iraq and directing operations in Afghanistan. Jasmine's awards include the General Douglas MacArthur Leadership Award, John M. King Award, and the Purple Heart.

The Great War's Far-Reaching Influence
for Black Male Actors Today

DARRYL DILLARD

WWI was the war to end all wars. Or so we thought. It came at a time when America was still dealing with its own struggle of equality in its own walls. To prove loyalty to the country and gain respect, blacks volunteered in droves to join the war and fight for their country side by side with whites. They wanted to be equals, willing to fight. They also believed this was their chance to be seen as just as human as white people.

Slavery had been officially abolished, but that did not stop the mistreatment of blacks. Abuse and lynching continued despite the new laws. Whites mostly thought that blacks were not humans, but animals, and less intelligent than whites.

The media of the time legitimized these thoughts and beliefs. Blacks in mainstream media were nonexistent except for white actors in black face. Al Jolson was the most prominent. His image was used for ads portraying blacks as animalistic, less human than whites.

This representation of blacks as animals continued in films such as D.W. Griffith's 1915 *Birth of a Nation*, in which the good Christian men of the community, tried to protect their community from the scourge of the black man. Dressed in white sheets they tracked down an aggressive black man, played by a white actor in blackface who was destroying their town and raping women to kill him. Another scene from the time has a white woman jumping off a cliff to keep from being taken by a black man. Again, the actor was white, in blackface.

These examples helped establish the idea that the black man must be kept down or killed. The only way for the black man to continue in film, according to the media, was to be entertaining and funny for white people. The shuck and jive type acts were popular. But again, instead of actual black actors, white actors in blackface performed the roles. They kept up with what they liked about, but only to the benefit of whites.

The image of animalistic blacks in the WWI era also appeared in propaganda. During WWI, to recruit more soldiers for the war, posters and campaigns portrayed gorillas as enemies. The media frequently used gorilla imagery to symbolize blacks. The Hun on propaganda posters, who raped Belgium women and stuck bayonets through babies, was a bad, black man.*

This propaganda contributes to Willie Lynch-type media issues we deal with today. Willie Lynch was a slave owner in Virginia who is said to have given a speech in Virginia in the 1700s that instructed masters to pit slaves against one another according to such criteria as age, skin color variation, ability, and sex to keep them on plantations instead of threatening them with lynching. Basically, to build mistrust and fear among the slaves to keep them physically strong, and, as a result, mentally weak.

This "Willie Lynch syndrome" has been perpetuated today in the black community regarding the military—a mentality exists that equates joining the military with a way of getting instant respect and honor for being a real, strong man. It creates what some might call the Willie Lynch-style competition. Damned if you do, damned if you don't.

To me, it's ironic to note that WWI posters portrayed the German as a black brute while blacks were allowed to fight alongside whites in some of the biggest battles of the war, such as the Meuse-Argonne. Even more troubling—they weren't even recognized until much later.

WWI-era racist images and mental sabotage persist in today's media. While they may appear subtle, they are very powerful, and easy to fall prey to if you do not know history. In 2008, Vogue published a cover photo of LeBron James holding Gisele Bündchen in a way that resembled the WWI posters "Destroy this Mad Brute - Enlist" in which an aggressive gorilla ravishes a blond, white woman. For me, it's not a coincidence that these images are almost identical. Vogue in 2008 seems to have turned a blind eye to the racist part of the 1917 poster, which is to my eye, obvious. Some say Vogue was benignly comparing LeBron James to King Kong, not an evil-German-gorilla-black man. But even then, King Kong films have strong racial overtones.

Looking at this poster makes me understand that perhaps we have not come so far from WWI in the portrayal of the black man in modern media. The black man is still feared by many and images that coincide with this fear justify prejudice.

That's why, as a black male actor, I decided to make and write the film, *The Real Man,* with Cathy Reinking, producer, director, writer, casting director, and coach. *The Real Man,* directed by Reinking, tells the story of a black actor who vows to only play uplifting roles. After discovering he can't make a living this way, he decides to break his own rules and accept negative stereotypes as well as one-dimensional parts. As he becomes successful with each new dark role, he begins to jeopardize the strong relationship he has with the love of his life, Laura, who is white.

As artists, our attention is now toward telling our own stories, which are personal and yet also examine media images and their power and influence in the world. It's easy to be swayed into playing someone you are not if your sense of self is not strong. And that's what happens to Donny, the lead character in my film. He gets lost in the stereotype characters the film Industry is paying him to be and it changes him for the worst.

In *The Real Man,* and in the other work I do as a writer and artist, I can't undo WWI imagery, but I do seek to counteract the history of negative representation of the black man in the media.

*See *Destroy This Mad Brute – Enlist* poster in essay by Ernest Lucas Mc-Clees, "Demonizing the Enemy: "One in the Same"" on page 163.

Darryl Dillard is an award-winning writer, actor, director, activist, producer, and casting associate. His credits include *The Real Man* (co-writer, co producer, and lead), *Then I'll Rest* (Actor, script consultant), *Who's with Me?* (Writer, director, lead), as well as network tv shows *Legacies* (Actor), *Valor* (Actor), *Star* (Actor), *The Resident* (Actor), *Ozark* (Actor), and an upcoming Starz show. He has worked the "In My Shoes" Spoken Word Event and Brunch as HOST and performer. He also co-wrote and acted in the play "Hardball," performed in Atlanta.

His latest projects as a director, *Cowboy Joe, Trauma-less Therapy*, and *Blindsided* are in festival circuits.

First I Said No: Writing about WWI African American Doctors in Fort George G. Meade's 100-Year History

MARY L. DOYLE, veteran

Fort George G. Meade held the first two-hour planning meeting for our WWI Centennial celebrations in August 2015. After the meeting, I headed toward the door with a long list of events, projects, and further planning to do. Camp Meade had been one of the first of sixteen cantonments established at the start of the U.S. involvement with WWI. The plans we made to acknowledge that history were going to keep me busy for the next couple of years and I couldn't wait to get started.

Then Barbara Taylor, the now-retired Fort George G. Meade museum curator, grabbed me by the arm and whisper-hissed her declaration to me.

"We have to write a book."

If you're a writer, you grow accustomed to people saying this to you. My first reaction was absolutely, positively no. N. O. Not gonna happen. Nope. Not me. Nein!

But Barbara persisted until I finally agreed to serve as principal editor and writer of *Fort George G. Meade: The First 100 Years.* The e-book and hardcover contain nearly thirty essays about WWI alone, which accounts for only two years of the installation's 100-year history. From WWI to the present, including the establishment of U.S. Cyber Command—the nation's newest combatant command that put its down roots at Fort Meade—the book provides an overview of it all.

I'd worked at Fort Meade in Maryland for over ten years at the time of the centennial planning. I'd read bits and pieces of the history and knew some of the legends. Named after the Civil War general, home of the first Tank Corps and the place where George S. Patton and Dwight D. Eisenhower developed tank tactics. Fort Meade had a museum where one of the first battle tanks used in WWI was housed. To that point, there had been a small pamphlet published about the history of the installation but not much else.

The only non-fiction books I'd written were co-authored memoirs of two amazing African American women. I'd helped write their memoirs because they'd lived lives that history needed to record urgently before their stories disappeared forever. Shoshana Johnson, the first African American woman to be held as a POW, and Brigadier General (Ret.) Julia Cleckley, the first African American women general of the line in the Army National Guard, had compelling, important accounts that reflected honest and raw experiences as the lives of black women in military uniform.

Yes, I'd co-authored two memoirs, but most of my writing is fiction. Even in fiction, my characters are women who wear combat boots, characters who, in genres like mystery, urban fantasy and erotic romance, work their way through life and death challenges while demonstrating bravery, intelligence, and sharp, biting humor. With stories based largely on my experiences in uniform, I use genre fiction to share the lives of my African American women characters who move through the male-dominated military space, facing the challenges presented by both their gender and their race.

After publishing seven books in fiction and non-fiction, I couldn't imagine how I would take on a project like Fort Meade's military history, filling it with dates, heavy details and long recitations of facts and figures. I didn't have the expertise and we didn't have the time.

Barbara made the argument, with breathless excitement, that we could make it a community project, get local historians and others with connections to the base to submit essays, to write about what interested them. She said she'd comb the installation and national archives; she'd do the research. She'd thought it all through. She'd planned her argument. She was determined to convince me to take on the project. All I had to do was write it.

One of the first essays we received cemented the value of the project to me. Joann Buckley and Douglas Fisher wrote an essay based on the research they did for their book, *African American Doctors of WWI: The Lives of 104 Volunteers*. In the essay, they tell the story of two universities—Howard University in Washington, DC and MeHarry Medical College in Nashville, Tennessee—that helped recruit doctors to join the Army, knowing black soldiers of the 92nd and 93rd Infantry Divisions wouldn't get proper medical care if African American doctors didn't provide it. Since a large component of the 92nd Division was at Camp Meade, eight of those doctors happened to be stationed there.

The African American doctors reported to Camp Meade after going through the hastily established officer's course developed for physicians at the segregated officer training camp at Fort Des Moines, Iowa. The eight doctors, Arthur L. Curtis, Thomas E. Jones, Oscar DeVaughn, Raymond W. Jackson, John H. Williams, James Wittico, William A. Harris, and William J. Howard along with the 110 other doctors stationed elsewhere, had left safe and prosperous medical practices to serve.

The stories of these brave men may have disappeared if not for the work of Buckley and Fisher.

Just after the U.S. declared their intention to enter WWI, many black men who tried to join were turned away. Eventually, black men were allowed to sign up but were restricted to labor and animal handling jobs. The 92nd and 93rd Infantry Divisions weren't formed until African Americans lobbied for the opportunity to fight and die for their country. After working at Fort Meade for a decade, I'd never heard the story about the black doctors, or the 92nd Infantry Division soldiers who trained on Camp Meade. I'd heard about the thousands of men who trained in the first tanks to be deployed in battle. I'd heard about the revolutionary medical care in prosthetics, the nurses who trained to operate in trenches, and I'd heard about the "Hello Girls," reportedly the first women to officially serve in uniform who performed trunk operator duties both in the U.S. and in France.

We don't know much about African American contributions because no one was recording it. In all our research, I never saw anything about the role black women played in the war at Camp Meade. At some point, black nurses were trained and allowed to serve, but the program began too late for any of them to get to Camp Meade. It's hard to believe that there weren't black women who stepped up, who rolled up their sleeves, and played notable roles. I can only guess that since their stories weren't recorded, those roles will forever go unnoticed and unrecorded.

I think it's funny now that I said no before I said yes to this project. There were times when I wished we were writing the story of 100 years of drinking "mead" and had plenty to sample. Other times, I wished I'd stuck with my first response. I'm still no historian, but now I have raging respect for the people who put on the white gloves and sit in dusty rooms digging through papers and books and connecting all the dots and telling us the stories. It's sad I didn't find stories about black women who made their contributions. Despite not finding their stories, I know they are there. Now, I must encourage women, especially African American women, to tell their stories. If they don't, 100 years from now another writer will be left guessing.

Mary L. Doyle is the co-author of, *I'm Still Standing; from Captured Soldier to Free Citizen, My Journey Home,* (Touchstone, 2010). Mary has also authored the award-winning Master Sergeant Lauren Harper mystery series, a four novella erotic romance series, two books in an urban fantasy series and served as the principal writer and editor of *Fort George G. Meade; The First 100 Years.* Mary served in the U.S. Army Reserve and, as a career Army Civilian. Her essay work has been published in *The Goodman Project*, the *War Horse*, and *O-Dark Thirty.*

War Isn't the Only Hell: 100 Years Later, Time to Tell the Truth about the African American and Lost Generation Experiences

KEITH GANDAL, PhD

Our sense of American Great War writing—and thus of the American experience—is limited and misguided. It may seem surprising, but, unlike every other combatant nation, our lasting WWI literature was written entirely by noncombatants. Hemingway was a Red Cross ambulance driver, as was Dos Passos, who later served in the army, but in the Services of Supply; Faulkner trained with the Canadian Royal Air Force but never made it over to Europe; Fitzgerald made junior officer in the American training camps, but his poor performance guaranteed that he was never shipped to France. By comparison, to take just the example of Britain, canonized authors Wilfred Owen, Siegfried Sassoon, and Robert Graves were all combat soldiers. Moreover, we don't know what it meant to be an American noncombatant male then, a status that shaped the unique writing of the Lost-Generation in ways many of us don't fully understand.

Serving as noncombatants was hardly what the Lost Generation young men had been counting on, given their social privilege; they were all expecting to be officers or pilots. But, surprisingly breaking with previous tradition, the army instituted a policy that gave equal opportunity to ethnic-American and working-class men, based on merit. In fact, that war involved a chapter of American history that eventually changed this country forever. WWI was America's first large-scale, national experiment in meritocracy, a conception of status and a practice of governance that had roots in the founding principle of human equality and that would move the country toward an understanding of equality not simply as equality of political rights but also as equality of socioeconomic rights, or equality of opportunity for advancement.

More than one hundred years after U.S. involvement in WWI, it is time to revisit American literature that came out of that conflict, because we are only now able to understand it in its historical context. That is the purpose of my new book, *War Isn't the Only Hell: A New Reading of World War I American Literature.*[1]

My interest in WWI and its literature was not sparked by my family but by my graduate work thirty-five years ago at Berkeley, with the French "post-structuralist" philosopher and historian, Michel Foucault, who was interested in modern forms of what he called "governmentality" and what we would call social control or organization. Back in 1983, I spent a month doing research in the U.S. Military archives. I discovered that the army had used all kinds of new-fangled "personnel" techniques to evaluate, assign, and promote recruits during World War I, including the notorious intelligence tests. I kept these papers for the next quarter century—packing them up and taking them with me each time I moved—because I felt they were important.

It wasn't until the mid-2000s that I returned to these military papers. In the intervening years, a new generation of social-military historians had done a good deal of groundbreaking research, which I also consulted in writing my first book on the subject: *The Gun and the Pen: Hemingway, Fitzgerald, Faulkner and the Fiction of Mobilization.*[2] *War Isn't the Only Hell* is a follow-up to that one, with a much broader reach. The unusual tale of this research and writing—which has transpired over the course of my entire career in academia—has also become the subject of my in-progress memoir.

In *War Isn't the Only Hell,* I've tried to expand the canon, beyond the work of famous Lost-Generation authors Hemingway, Faulkner, Fitzgerald, Dos Passos, Cummings, and Katherine Anne Porter. American WWI literature, by combatants and noncombatants alike, has long been interpreted as an alienated outcry against modern warfare and propaganda. But this traditional understanding ignores the U.S. army's unprecedented attempt, during the war, to assign men—except, notoriously, African Americans—to positions and ranks based on merit. In the Civil War, state governors handed out officer commissions, as political favors, based on family and social status.

The conventional reading also misses the fact that the army and the culture granted masculinity only to combatants, while the noncombatant majority of doughboys experienced a different alienation: the shame of rejection and emasculation. The army did not set out to alienate what turned out to be three-quarters of its recruits, but it was essentially a victim of its own success in selling the war and the draft with traditional images of warriors as combatants. Wartime propaganda, as well as the tales expected by families back home, left many noncombatants feeling less than masculine. Meanwhile, many of the men chosen to be combatants experienced a sense of affirmation or promotion.

This partly meritocratic army policy—which excluded black soldiers—was not a matter of social justice for immigrant minorities and the white working class, but of the practical, bottom-line aim of winning the war. Indeed, shocking though it might seem to us, given what we think we know about the WWI period, the army's original plan was for no racial discrimination against African Americans. The plan bowed to white fears in only one, namely that, at all training camps, white troops would overwhelmingly outnumber black ones.

To be sure, this plan did not go into effect, and the reason was that the military brass was overruled by a federal government pressured by alarmed white southern leaders. They feared such a plan would undermine white supremacy. But the army did not simply cave into white southern demands. The ultimate military policy adopted concerning African Americans did not please black leaders, but it displeased white racist leaders too. Some high-profile white southern civilian leaders wanted no African Americans drafted or sent to France, and none wanted black commissioned officers, black troops trained in combat, or black recruits stationed in southern states. The military, though it acquiesced in some ways—most notably in instituting segregation and assigning most black recruits to labor units—insisted on a black draft, commissioned about 1,200 black officers, put tens of thousands of black troops into combat, and did station black recruits in southern states. Perhaps most significant for the African American community, the army gave hundreds of thousands of black men a chance to experience, France, a country not governed by racial discrimination who embraced them.

Drawing on military archives and current historical research, my book discusses the work of thirteen significant writers: as responses to the shocks of war and meritocracy. The supposedly antiwar texts of the social-privileged male Lost-Generation authors addressed—often in coded ways—noncombatant frustrations. Meanwhile, the hard-hitting works of combat soldiers William March, Thomas Boyd, Laurence Stallings, and Hervey Allen were partly shaped by experiences of meritocratic recognition, especially meaningful for socially disadvantaged men.

Even the sole WW I novel by an African American veteran, Victor Daly, reveals a mixed experience of discrimination by the U.S. Army and a feeling of empowerment from the French. Finally, some of our best-known WWI women writers—Porter, Willa Cather, and Ellen La Motte—were reacting not only to the shock of modern war, but to new employment and social opportunities for women at the front and stateside, including their new power, at home, as arbiters of masculinity, a prerogative that temporarily altered gender relations. Ultimately, this literature registered the ways in which innovative military practices and a foreign war unsettled traditional American hierarchies of class, ethnicity, gender, and even race.

Most of my book is dedicated to the lesser-known writers mentioned here. Its title comes from Daly, who covertly tells a taboo story, most of which has been missed by critics, of a black soldier who sleeps with an upper-class and cultured white French woman.

Since we've passed the WWI centennial, I suggest we broaden our canon so this long-forgotten American war becomes a major cultural touchstone.

Keith Gandal is Professor of English at City College of New York, with a joint appointment in American literature and creative writing. His publications have had three focuses: urban poverty, war and mobilization, and modern medicine and illness. He is the author of four scholarly books: *War Isn't the Only Hell: A New Reading of World War I American Literature* (Johns Hopkins University Press, 2018); *The Gun and the Pen: Hemingway, Fitzgerald, Faulkner and the Fiction of Mobilization* (Oxford UP, 2008); *Class Representation in Modern Fiction and Film* (Palgrave Macmillan, 2007), and *The Virtues of the Vicious: Jacob Riis, Stephen Crane and the Spectacle of the Slum* (Oxford UP, 1997). He is also the author of a novel: *Cleveland Anonymous* (North Atlantic Books, 2002).

Notes

[1] Keith Gandal, *War Isn't the Only Hell: A New Reading of World War I American Literature* (Baltimore: Johns Hopkins University Press, 2018).

[2] Gandal, *The Gun and the Pen: Hemingway, Fitzgerald, Faulkner and the Fiction of Mobilization* (New York: Oxford University Press, 2008)

Their Only Crime: African American WWI Poet James Seamon Cotter, Jr.

CONSTANCE M. RUZICH, PhD

Although a citizen of the United States, the black man is regarded by the white American as an inferior being with whom relations of business or service only are possible. The black is constantly being censured for his want of intelligence and discretion, his lack of civic and professional conscience and for his tendency toward undue familiarity. The vices of the Negro are a constant menace to the American who has to repress them sternly.[1]

–"Secret Information Concerning Black American Troops," sent August 7, 1918, from AEF Colonel J.L.A. Linard to the French Army

Over 350,000 black Americans were inducted into the American Army during WWI, but units were strictly segregated by race, and black soldiers were assigned to hard labor and low status jobs such as the grave digging, exhumation, and reburial work of the war. Few black units saw combat; units assigned to the French military provided an exception and they fought with bravery and distinction. In the WWI American Army, racism was not only accepted but often enforced.

James Seamon Cotter, Jr. has been described as a "forerunner of the African American cultural renaissance of the 1920s,"[2] and the *Encyclopedia of the Harlem Renaissance* notes that his poetry and one-act play *On the Fields of France* provide an important contribution to WWI literature. Cotter's poem "O Little David, Play on Your Harp" uses a well-known African-American spiritual to frame the oppression and misery of war, genocide, and racism.[3] In 1919, Lt. Noble Sissle and Lt. James Reese Europe of the 369th Harlem Hellfighters recorded their performance of the song:

O Little David, Play on Your Harp

O Little David, play on your harp,
That ivory harp with the golden strings
And sing as you did in Jewry Land,

Of the Prince of Peace and the God of Love
And the Coming Christ Immanuel.
O Little David, play on your harp.

> A seething world is gone stark mad;
> And is drunk with the blood,
> Gorged with the flesh,
> Blinded with the ashes
> Of her millions of dead.
> From out it all and over all

There stands, years old and fully grown,
A monster in the guise of man.
He is of war and not of war;
Born in peace,
Nurtured in arrogant pride and greed,
World-creature is he and native to no land.
And war itself is merciful
When measured by his deeds.
Beneath the Crescent
Lie a people maimed;
Their only sin—
That they worship God.
On Russia's steppes
Is a race in tears;
Their one offense—
That they would be themselves.
On Flanders plains
Is a nation raped;
A bleeding gift

> Of "Kultur's" conquering creed.
> And in every land
> Are black folk scourged;

Their only crime—
That they dare be men.
O Little David, play on your harp,
That ivory harp with the golden strings;
And psalm anew your songs of Peace,
Of the soothing calm of a Brotherly Love,
And the saving grace of a Mighty God.
O Little David, play on your harp.[4]

The celebratory refrain of the Negro spiritual contrasts sharply with a "seething world" that has "gone stark mad." Spiraling out of control, the world at war is drunk on blood, sated by the decaying bodies of the dead, and blinded by the ashes of destruction. Yet bigger than the war and more terrible than its slaughter, a monster "of war and not of war" towers overall. This fiend, born in peace, raised by pride, and fed by greed, is a citizen of every nation, and he wears a human disguise. In the Ottoman Empire—"beneath the Crescent"—he has directed the massacre of the Armenians; in Russia's pogroms; he has murdered thousands of Jews; and he has brutally commanded German atrocities in occupied Belgium. Cotter's poem unites these victims of deadly prejudice with blacks who are whipped and beaten "in every land;" their only crime is daring to believe themselves fully human.

Many black Americans hoped the war that was to "make the world safe for democracy" would also address the racism that was prevalent in America. In "O Little David," Cotter challenges his audience to acknowledge that the enemy within, the "monster in the guise of man," is as terrible a foe as any to be encountered on the battlefields of Europe.

Constance M. Ruzich was a 2014-2015 Fulbright Scholar at the University of Exeter, where she researched the use of poetry in British centenary commemorations of WW1. She is the editor of *International Poetry of the First World War: An Anthology of Lost Voices* (Bloomsbury, 2020), and she runs the popular blog *Behind Their Lines*, which discusses poetry of the Great War. Her essay "Distanced, Disembodied, and Detached: Women's Poetry of the First World War" appears in *An International Rediscovery of World War One: Distant Fronts* (Routledge, 2020). She is a professor of English at Robert Morris University in Pennsylvania.

Notes

[1] Colonel J.L.A. Linard to French Military Mission, "Secret Information Concerning Black American Troops," August 7, 1918. In "Documents of the War," collected by W.E.B. DuBois, *The Crisis* 18, no. 1, (May 1919): 16-18.

[2] James Robert Payne, "Joseph Seamon Cotter, Jr.," in *The Concise Oxford Companion to African American Literature,* ed. William L. Andrews, Frances Smith Foster, and Trudier Harris (Oxford, UK: Oxford UP, 2001): 90.

[3] The subject of the song, however, is relevant to the poem's message. David's harp playing was commanded by King Saul, who employed the boy to soothe his mad rages (I Samuel 16), and the young shepherd shocked Israel's army with his courage and skill in fighting the colossal Goliath (I Samuel 17).

[4] Joseph Seamon Cotter, Jr., "O Little David, Play on Your Harp," in *The Band of Gideon and Other Lyrics* (Boston: Cornhill, 1918), 15–16.

The Story of Freddie Stowers, the First African American Recipient of the WWI Medal of Honor*

COURTNEY L. TOLLISON HARTNESS, PhD

The centennial commemoration of the end of WWI was a long-anticipated global event. Around the world, throughout the years and months leading up to November 11, 2018, the centennial of major offensives and battles, and the deaths of millions who died in them, was observed in meaningful and elaborate ceremonies.

Over the course of my career, I have been fortunate to have spent time interviewing people about their lives and their contributions to historical events. From the living, we also learn about those who died and never got to tell their story. From 2007 to 2010, my students and I interviewed over sixty men and women who served in, contributed to, and/or were personally affected by World War II. A significant number of these interviewees were African American veterans. Their oral histories reflect courage and pride but were often tinged with bitterness and resentment resulting from the way African Americans were treated during that era. After Barack Obama's election to the presidency, it was fascinating to hear how these veterans' perspectives on their service had evolved.

A few years later, I was privileged to spend two weeks participating in a seminar at the U.S. Army Command and General Staff College at Fort Leavenworth, Kansas. At their thirteen-foot bronze monument to the Buffalo Soldiers, dedicated in 1992 by General Colin Powell, I reflected on the African American veterans I had interviewed, and how the complex feelings about their service during World War II must have been shared by Buffalo Soldiers who preceded them.

In anticipation of the centennial of WWI, I began researching African Americans who served in the American Expeditionary Forces (AEF) from my community. Most compelling was the story of one soldier whose influential and enduring legacy would have been inconceivable to him during his lifetime. September 28, 2018, marked the centennial of his death.

Freddie Stowers was born in the tiny town of Sandy Springs in upcountry South Carolina in 1896, the same year that *Plessy v. Ferguson* provided judicial justification for racial segregation. As the grandson of slaves, racial barriers of the day would have circumscribed Stowers' life in this small southern town. He married a young woman named Pearl. They named their daughter, Minnie, for one of his sisters. He worked as a farmhand before being drafted into the Army, where he served with the 371st Infantry Regiment of the 93rd Division, one of only two African American divisions. Before sailing to France, he trained at Camp Jackson in Columbia, South Carolina, about two hours away from his hometown.

As an African American infantryman in WWI, Stowers was an extreme minority. In 1917, when the U.S. entered the war, our country was only fifty-two years removed from the Civil War that had broken us apart over racial issues. High racial tensions prompted many white southerners to fear the arming and training of mass numbers of African American males, and white southern politicians lobbied against using them in combat and placing them in southern training camps. Ultimately, racial prejudice dictated the roles and responsibilities approximately ninety percent of African American males held during the war, which included loading and unloading cargo, digging trenches, burying bodies, cooking, and chauffeuring.

Stowers was among the ten percent of African American servicemen from WWI who saw combat.

As a member of the 93rd Division, Stowers served under French command. Only two days after American Expeditionary Forces launched the Meuse-Argonne offensive, Stowers led an attack on Hill 188, which had been fortified by the Germans. As Stowers and his fellow soldiers approached, enemy troops climbed out of their trenches with their hands up in supposed surrender. Yet when Stowers and the others drew closer, the German troops opened fire. Stowers pressed onwards to the machine gun nest that posed the greatest threat, and, after those senior to him in rank had fallen, he encouraged his men to similarly attack the enemy's defensive fighting position. After successfully overcoming the first set of trenches, Stowers reorganized the men in his platoon and attacked the second set of trenches. During this attack, Stowers was struck but fought on until struck again. As he collapsed, he continued to encourage his men to forge ahead. They were successful in driving the Germans from the hill. Stowers died that day, at the age of twenty-two, and is buried in the Meuse-Argonne American Cemetery about 160 miles northeast of Paris.

While the circumstances of his service and sacrifice are enough to warrant our gratitude and respect, the continued impact of his death in redressing racial inequities in our country has helped transform our nation. The entire 93rd regiment was awarded France's Croix de Guerre (War Cross), an esteemed military honor. At the time, deeply rooted racial prejudice infiltrated nearly all aspects of American society, and the United States government failed to recognize the efforts of its African American soldiers.

By 1990, not a single African American soldier had received the Medal of Honor for WWI or WWII. That year, the U.S. Congress asked the Army to review Medal of Honor records, citing racial inequities in awarding of medals from the world wars. The United States' highest military honor, the Medal of Honor, originated during the Civil War. African American soldiers received this distinction for service in not only the Civil War, but also Native American Wars, Spanish American War, and the Korean War in the 1950s.

During their investigation, the Army located paperwork from WWI recommending Freddie Stowers for a Medal of Honor. A team traveled to France to study the circumstances of his death, and the Army Decorations Board approved the honor.

Seventy-three years after he was killed in action, Freddie Stowers became the first African American recipient of the Medal of Honor for WWI. Stowers' surviving sisters, Georgina and Mary, stepped forward at the White House ceremony on April 24, 1991, to accept his medal from President George H. W. Bush.

The posthumous awarding of the Medal to Stowers contributed to an official report that decried institutional racism as the root cause for the lack of military honors to African American soldiers in both world wars. Furthermore, subsequent studies suggested that discrimination had impacted not just African American soldiers, but also Jewish Americans, Hispanic Americans, and Japanese Americans.

Presidents Clinton, George W. Bush, and Obama continued the commitment begun by President George H.W. Bush to redress these inequities. Collectively, from 1991 to 2015, these presidents awarded over fifty Medals of Honor, most of them posthumously to male minorities for their service in World Wars I and II, Korea, and Vietnam.

Given the racial climate within which Stowers lived, the lasting and powerful impact of his courageous service on our nation's treatment of our most lauded men and women in uniform would have been unfathomable to him. When Stowers was a child, President Teddy Roosevelt shocked the nation when he dined with African American, Booker T. Washington, at the White House; ninety years later, the White House hosted a ceremony in his honor.

Stowers' death contributed to the campaign to end a nasty war. One hundred years later, his legacy has contributed to the mitigation and reconciliation of our country's long history of racial prejudice.

* A version of this piece was first published in the *Greenville News* on September 30, 2018.

Courtney L. Tollison Hartness, PhD is the Distinguished University Public Historian and Scholar at Furman University. She has been a Fulbright Scholar (Ukraine), a fellow in Columbia University's Oral History Research Office, and a Mellon Grant recipient for ASIANetwork's Faculty Enhancement Program (India). She is also a graduate of the U.S. Army Command and General Staff College Military History Instructors Program at Fort Leavenworth. She has published two books, curated four museum exhibits, worked on several documentaries, conducted over 100 oral histories, and served as historian for memorials, markers, and sculptures throughout South Carolina. Her current research focuses on South Carolina during WWI and the Progressive Era.

part five

BRAVERY AND RESILIENCE: NATIVE AMERICANS IN WWI

Writing the Story of California's
Muwekma Ohlone Indian Tribe in WWI

BY ALAN LEVENTHAL, PhD

I was hired at San Jose State University in the Department of Anthropology in 1978. As a trained archaeologist, I worked on a multitude of ancestral Ohlone heritage sites. However, at that time, there was scant information about the Ohlone Indians—also referred to as the Costanoan Indians—of the San Francisco Bay region. In 1980, two anthropology students escorted Chairwoman Rosemary Cambra, who claimed to be an Ohlone Indian to my office. After having an engaging discussion about her family and her research goals relative to her tribe, I informed Rosemary that, although I had taught about Native American tribes at the American Museum of Natural History and worked directly with Washoe, Paiute, and Shoshone tribal communities in Nevada, I knew virtually nothing about California Indians, let alone about the history and heritage of the Ohlone/Costanoan tribal groups.

I then took Rosemary to our university library—before computers—and we used index cards in oak file drawers. We found Ohlone Indians on a card, which also noted for us to "see Costanoan." The monumental tome, *Handbook of Indians of California* written by eminent California anthropologist Alfred L. Kroeber from University of California at Berkeley was the most comprehensive source we found.[1] Published in 1925 as "Bulletin 78" of the Bureau of American Ethnology of the Smithsonian Institution, Kroeber's handbook became the quintessential authoritative publication on all of the California Indian tribes.

In his chapter on "The Costanoans," Kroeber writes, "The Costanoan group is extinct so far as all practical purposes are concerned. A few scattered individuals survive, whose parents were attached to the missions of San Jose, San Juan Bautista, and San Carlos; but they are of mixed tribal ancestry and live almost lost among other Indians or obscure Mexicans."[2]

After reading this section, I turned to Rosemary and said to her that she must be from some other tribe because Kroeber had declared the entire group extinct by 1925. Rosemary responded that she was an Ohlone Indian, her mother was born on the Sunol rancheria in the East Bay, and she was baptized at Mission San Jose in 1912 with her other siblings. So rather than take issue, I asked her what research she wanted to explore. Rosemary stated that her family, extended families, and other non-related families wanted to explore the mission baptismal and death records to see if they could trace their respective genealogies back to their aboriginal villages. I agreed to work with those families who were interested in conducting such research.

As I interviewed various elders of the tribe, I came to realize that much of their twentieth-century history was not included in Kroeber's handbook, including the fact that they were federally recognized in 1906 as the Verona Band of Alameda County. During interviews, the elders remembered men coming to their houses when they were children in the 1920s and 1930s, asking about their spoken Indian language. I would later find out that the linguist who visited them was John Peabody Harrington from the Bureau of American Ethnology, Smithsonian Institution.

Further discussions centered around their enrollment with the Bureau of Indian Affairs in the following year groups: 1928-1932, 1948-1957, 1968-1971. They also discussed attendance in Indian Boarding Schools such as Sherman Institute in southern California and Chemawa Indian School in Salem, Oregon during the 1930s and 1940s. They explained their dynamic, rich history of service in the United States Armed Forces during WWII in multiple divisions and units.

After interviewing some WWII veterans, they pointed out that several men of their parent's generation had served in WWI. Working with various sources, we were able to track down their service records and burial locations in the Golden Gate and Riverside National Cemeteries. I also came to realize that not one military historian included their service in the United States Armed Forces spanning from WWI, WWII, Korea, Vietnam, Desert Storm and the recent conflict in Iraq. This includes not only the men and women of the Muwekma Ohlone Tribe but also those from the 135 California tribes and bands that were illegally removed from their Federally Acknowledged status in 1927 by an incompetent Superintendent at Sacramento, Lafayette A. Dorrington.

A CALL TO WAR: MUWEKMA MEN ENLIST IN THE U.S. ARMED FORCES DURING WWI

The following paragraphs present those Muwekma men who served in the Army, Navy and Marine Corps during WWI, some even before America's official involvement in 1917. Even before California Indians legally became citizens in 1924, at least six Muwekma men joined 17,000 other Native Americans and served in the United States Armed Forces in the Army, Navy, and Marine Corps. These Muwekma men enlisted through the San Francisco Presidio and Mare Island and four of them are buried at the Golden Gate National Cemetery.

ANTONIO (TONEY) GUZMAN

Toney Guzman was born on March 27, 1890, either in Centerville or on the Niles Rancheria. He was the son of Muwekma Indians Francisca Nonessa and Jose Guzman. Toney enlisted in the U.S. Army and fought in the Meuse-Argonne from September 26 to October 8, 1918, Ypres-Lys, and Lorraine campaigns in France. Toney served in the Army from April 29, 1918 and was honorably discharged at the San Francisco Presidio on April 26, 1919.

The 91st Division was known as the "Wild West Division." The Division's shoulder patch represented a green fir tree referring to its origin at Camp Lewis in the Pacific Northwest. The Division deployed to France in August 1918 and fought with great distinction. In the Ypres-Lys campaign, the Division served in the Flanders Army Group, under the command of the King of Belgium. The Division was headquartered adjacent to Flanders Field. Five members of the Division earned the Medal of Honor. The 347th Field Artillery Regiment was assigned 4.7"-inch guns, and the 91st Division received the following Victory Medal Clasps: Ypres-Lys, St. Mihiel, Meuse-Argonne and Defensive Sector.

In October 1931, Toney Guzman and his brothers enrolled with the Bureau of Indian Affairs under their mother's BIA Application #10293. On his WWII Registration Card dated April 27, 1942, Toney was identified as "Indian." Toney died on October 8, 1948, and was buried on October 12, 1948, at the Golden Gate National Cemetery (Section J, Grave 254).

ALFRED (FRED) GUZMAN

Alfred Guzman was born on the Pleasanton Rancheria on June 27, 1896, to Francisca and Jose Guzman. Prior to the declaration of WWI, Fred Guzman had served in the National Guard at Fort Mason in San Francisco in 1917. Afterwards he enlisted in the U.S. Army, and served in the 28th Division, 55th Brigade Infantry, 110th Infantry, Company "C" and fought in the major battles at Ourcq-Vesle (July 28, 1918), Second Battle of the Marne (July 15 to August 5, 1918), Meuse-Argonne Offensive (September 26 to October 8, 1918), and Havrincourt (October 8 to November 11, 1918) in France.

The 28th Division fought in the following campaigns: Champagne-Marne, Aisne-Marne, Oise-Aisne, Meuse-Argonne, Champagne (1918), Lorraine (1918). The cost in lives of these six campaigns was 4,183 casualties including 760 dead. The six fleurs-de-lis on the regimental insignia commemorated their World War I service. The 28th Infantry Division was a unit of the United States Army formed in 1917 at the outbreak of WWI. It was nicknamed the "Keystone Division," as it was formed from units of the Pennsylvania Army National Guard; Pennsylvania is known as the "Keystone State." It was also nicknamed the "Bloody Bucket" division by German forces in WWII, after its red insignia. Fred Guzman served from July 28, 1917 and was honorably discharged at San Francisco Presidio on May 31, 1919. On his WWII Registration Card dated April 25, 1942, Fred is identified as Indian. Fred Guzman died on November 3, 1961 and was buried at the Golden Gate National Cemetery (Section Y, Grave 1059).

JOSEPH ALEAS

Joseph Aleas was born on the Alisal (Pleasanton) Rancheria on May 11, 1893 and was the son of Margaret Armija. He enlisted in the U.S. Army on June 30, 1916. According to Armija-Thompson family recollections, he was a good horseman and wanted to fight against Pancho Villa had led approximately 1,500 Mexican raiders in a cross-border attack against Columbus, New Mexico, in response to the U.S. government's official recognition of the Carranza regime. Villa's troops attacked a detachment of the 13th U.S. Cavalry, seized 100 horses and mules, burned the town, killed ten soldiers and eight of its residents, and made off with ammunition and weapons. President Woodrow Wilson responded by sending 6,000 troops under General John J. Pershing to Mexico to pursue Pancho Villa and his troops. This military mobilization was called the Punitive or Pancho Villa Expedition.

Later, Joseph Aleas served in France in the 21st Machine Gun Battalion, 7th Division whose Hourglass insignia dates to 1918. Organized originally to serve in the American Expeditionary Forces during WWI, the U.S. Army's 7th Infantry Division was created at Camp Wheeler, Georgia on December 6, 1917, and it fought in Alsace-Lorraine, France. The division also served as an occupation force in the post-war period. On October 10-11, 1918 the 7th was shelled for the first time and later it encountered gas attacks in the Saint-Mihiel woods. Defensive occupation of this sector continued from October 10th to November 9th during which the infantry regiments of the 7th Division probed up toward Prény near the Moselle River, captured Hills 323 and 310, and drove the Germans out of the Bois-du Trou-de-la-Haie salient. After thirty-three days in the line of fire, the 7th Division had suffered 1,988 casualties, of which three were prisoners of war. Thirty Distinguished Service Crosses were awarded to members of the 7th Division.

Joseph Aleas was honorably discharged at Camp Funston, Riley, Kansas on July 9, 1920, and was awarded the WWI Victory Medal and the Bronze Victory Button. Joseph Aleas enrolled with the Bureau of Indian Affairs in October 1931 (BIA Application # 10299). On May 24, 1955, Joseph enrolled during the second enrollment period with the Bureau of Indian Affairs. Joseph Francis Aleas passed away July 13, 1964, and was buried at the Gold Gate National Cemetery Plot Z, grave 2597.

JOHN MICHAEL NICHOLS

John Michael Nichols was the older brother of Henry Nichols (see below) and he served in the U.S. Army from 1914 to1920. John enlisted on October 27, 1914, at Fort McDowell on Angel Island. He fought in France serving with the 59th Coast Artillery Corps which was attached to the 32nd Brigade, C.A.C. The 59th was converted to a TD (towed artillery) battalion and was engaged in the St. Mihiel offensive and the Meuse-Argonne offensive. John's artillery unit was later attached to Battery C, 67th Field Artillery, 42nd Infantry Division, at the time when he returned to the United States with his unit. John was discharged at Fort Winfield Scott at the SF Presidio on June 4, 1920. John M. Nichols was listed as an Indian on the 1930 Federal Census along with his son Alfred in Santa Cruz

County. On John Nichols's Draft Registration Card dated April 27, 1942, he was identified as residing at the Veterans Home in Napa in Yountville, California where he had been from 1941 to 1953. John Nichols died in April 1968 in Stockton, California.

HENRY ABRAHAM LINCOLN NICHOLS

Henry Nichols was born in Niles on February 12, 1895, to Charles Nichols and Muwekma Ohlone Susanna Flores Nichols. Henry enlisted on May 23, 1917, and first served on the USS *Albatross*. By December 31, 1917, he was transferred to the Battleship USS *Arizona*, and later March 26, 1918, he was transferred again to the Battleship USS *Oklahoma*. During WWI Henry Nichols served in the North Atlantic and was on escort duty in December 1918 when the Oklahoma escorted President Woodrow Wilson's arrival in France at the end of the war on November 11, 1918. The Oklahoma returned to Brest, France on June 15, 1919, to escort home President Wilson who was transported on the USS *George Washington* from his second visit to France. Henry Nichols was honorably discharged at Mare Island on August 14, 1919 and was issued the WWI Victory Medal. On Henry Nichols Draft Registration Card dated April 27, 1942, he is identified as Indian. Henry Nichols passed away on January 5, 1956 and was buried at the Golden Gate National Cemetery (Section L-5, Grave 7455).

FRANKLIN P. GUZMAN

Franklin P. Guzman was born on the Alisal Rancheria on January 15, 1898 and was the son of Muwekma Ohlone Indians Teresa Davis and Ben Guzman. He was also the nephew of Toney and Fred Guzman. Franklin was listed on the 1910 Federal Indian Population Census for "Indian Town," Pleasanton Township. Franklin enlisted with the local National Guard and was assigned to Company I, 5th Infantry Regiment, Second Brigade, in Livermore, Alameda County. It was mustered into Federal Service on June 28, 1916, as a response to the Pancho Villa incursions into the United States. He was discharged on October 7, 1916 and enlisted in the Marine Corps on October 20, 1916. While working near Sacramento, Franklin reported for duty on October 25, 1916, and was assigned to Company "B" Marine Barracks, Navy Yard, Mare Island. By March 31, 1918, he earned an Expert Rifleman Badge and a Marksman Badge. By April he was assigned to the 111th Company, 8th Regiment. In May, Franklin was transferred to the 150th Company 1st Machine Gun Replacement Battalion at Quantico, Virginia and was promoted to Sergeant on May 22, 1918. The 1st Machine Gun Replacement Battalion sailed on May 26, 1918, on the USS *Henderson* and disembarked in France on June 8, 1918. The 1st Machine Gun Battalion was later renamed the 6th Machine Gun Battalion in France. From September 12 to 16, 1918 the brigade was engaged in the St. Mihiel offensive in the vicinity of Remenauville, Thiaucourt, Xammes, and Jaulny. On September 16, 1918, Franklin Guzman was wounded in the left thigh and from September through December he was placed in various Field and Base Hospitals in France, and finally transferred back to the States on December 16, 1918. Franklin remained in recovery at the U.S. Navy Hospital at Norfolk, Virginia until he was honorably discharged from service as a Sergeant on June 27, 1919. Franklin's Battalion participated in the following battles: Chateau-Thierry sector for the capture of Hill 142, Bouresches, Belleau Wood from June to July 1918; Aisne-Marne (Soissons) offensive from July 18 to July 19, 1918; Marbache sector, near Pont-a-Mousson on the Moselle River from August 9 to August 16, 1918; St. Mihiel from September 12 to September 16, 1918; and later the Meuse-Argonne offensive from October 1 to 10, 1918, and from November 1 to 10, 1918. Franklin passed away on May 30, 1979 and was buried in the Riverside National Cemetery (Section 8, Grave 2826).

Alan Leventhal is a trained archaeologist/ethnohistorian. He worked at the American Museum of Natural History, departments of Anthropology and Education. Alan studied anthropology at City College of New York and San Jose State University (SJSU). In 1978, Alan was hired at SJSU and has co-authored numerous publications on Bay Area prehistory, California tribal ethnohistory and military service. He has served as a tribal archaeologist/ethnohistorian for the Muwekma Ohlone Tribe. In 2019, Alan retired from SJSU as staff in the Dean's office, College of Social Sciences, and since 1978 has taught classes about contemporary Native American Issues, and topics in archaeology.

Notes

[1] Alfred L. Kroeber, "Bulletin 78," in *Handbook of the Indians of California* (Government Printing Office, 1925), p. 464.

[2] Kroeber, "Bulletin 78," 464.

Soldiers Unknown

CHAG LOWRY (YUROK/MAIDU/ACHUMAWI)
Illustrations by RAHSAN EKEDAL

Cover, Soldiers Unknown. Used with permission of the author. Art by Rahsan Ekedal.

"Father, where did they go, and what did they see, these Yurok Native men who fought in World War One?"

"The answers are there for us to find, son. But the more important questions are how did they return home, and when did they find peace?"[1]

Crafting a World War One story with the very talented artist Rahsan Ekedal has been an emotional journey. We've come a long, long way since I first discussed the concept with him in 2016. I was raised among many Native veterans from both sides of my family in northern California. I always remembered the emotions of these men—many were WWII and Korean War veterans—when they told me about their fathers or uncles who had served in WWI.

Work in progress art from Soldiers Unknown *used with permission of the author. Art by Rahsan Ekedal.*

And that's the story Rahsan and I try to tell—a story about emotions. As a Native American, I wanted to create a graphic novel that conveys the experiences of the Great War through the sentiments of my people. I had two Yurok great-great uncles who served in WWI, and I've looked at their sepia-toned photographs for years wondering what they saw and felt.

The snippet of dialogue I shared at the beginning of this essay encompasses the questions I've contemplated regarding Native American veterans of WW1. Our story begins in contemporary times with a Father and a Son as they begin to talk about their ancestor's involvement in the Great War.

Can culture help a combat veteran find peace and return to their homeland? That's another question I thought about, and it's one I hope readers will discuss after they see our work. We used the lens of Yurok culture for this story to try to find an answer.

The beautiful part of being able to work with a respectful, talented artist like Rahsan is that his images allow me to feel the full range of emotions as a descendant of WW1 veterans. To teach young people about history we must first help them experience emotions and after that lead them to link their feelings to the lesson. The graphic novel, with its sequential art and impactful dialogue, can do just that. The act of breathing life into our characters in this story is meant to honor all WW1 veterans and their families. Rahsan fully colored his work, which means our soldiers will be as alive today as in 1917-18. We can share some of their journeys, and it doesn't matter what tribe or culture they are from. We can feel who they are. To most people, all WWI veterans are *Soldiers Unknown*. I hope this work helps change this.

Work in progress art from Soldiers Unknown used with per-
mission of the author. Art by Rahsan Ekedal.

Chag Lowry is of Yurok, Maidu, and Achumawi Indigenous ancestry from northern California. He created the Original Voices imprint to produce comic stories based on his people's history and perspective. He is the author of several books on Native American veterans and has produced and directed multiple PBS documentaries. He and artist Rahsan Ekedal crafted *Soldiers Unknown* over the course of three years. He's very proud that *Soldiers Unknown* is endorsed by the U.S. World War One Centennial Commission and the Yurok Tribe. Chag can be reached at ova4@sonic.net and on Instagram, Twitter, or Facebook.

Notes

1 Chag Lowry and Rahsan Ekedal, *Soldiers Unknown* (Pechanga, CA: Great Oak Press, 2019).

part six

FROM THE OTHER SIDE
OF NO MAN'S LAND

Remembering and Forgetting
the Great War in Germany

DAVID EISLER, PhD, veteran

They originally called it "Heroes' Grove" when they buried the dead from the Great War. In the lush, green hills of Germany's Rhein-Neckar Valley, about thirty kilometers from Heidelberg, the names of the town's forty-two fallen soldiers were engraved on individual plaques and attached to the oak trees that formed the woods. The families of the dead tended the graves.

In 1957, nearly forty years after the dedication of the Heroes' Grove, the plaques were recast in bronze and the names of the dead and missing from the Second World War were mixed with those of the First. A sandstone boulder with the years 1914-1918 and 1939-1945 bolted on its face was placed in the middle of the grove. Later, after nearly another forty years, the community would build two small stone walls near the central boulder and place the plaques with the names on them, where they remain today.

In 1998, eighty years after the end of the Great War, the local mayor encouraged the community to repurpose the land and expand the war memorial into a nationwide cemetery, preserving the peaceful atmosphere of the natural woods but rebranding the location as a burial destination for anyone. Ten years later, the community inaugurated the new Naturfriedhof ("natural cemetery") and changed the name from "Heroes' Grove" (Heldenhain) to "Resting Grove under the Oaks" (Ruhehain unter den Eichen). Today, for a few hundred euros, anyone can have a biodegradable urn with their ashes buried in the grove, with prices that vary depending on if you prefer to be next to a small boulder or a tree. Older, bigger trees are more expensive than little ones, which are more expensive than stumps.

World War I sits at a complicated place in German collective memory. Like the original Heroes' Grove, the memory retains a local connection, the towns that remember and mourn their dead without draping them in the war's politics—small plaques here and there with handfuls of engraved names that, spread over an entire country, lead to an undeniable sense of loss. That loss is compounded by the way Germans associate the First World War with the Second, a blending reflected even in the Ruhehain's two stone-bordered panels, just behind the large memorial stone, listing the names of those killed in both wars, mixed together with no obvious order or rationale; a subtle reminder that, here in Germany, the conflicts have combined into a single dark stain on the nation's collective memory, and of the challenge of remembering the fallen without honoring the cause for which they fell.

The site's transformation from a memorial for war dead to an eco-friendly burial destination for nature lovers raises interesting questions about the preservation of the war's collective memory. Historical scholarship has revisited and revised Germany's role in starting the Great War, and in the decades since the end of the Second World War, Germany has become wary of anything that sounds too much like militarism or nationalism. Unlike the Great War in British, French, or even American memory, the German way of memorializing the events tends to be more somber and understated, as though it would be wrong to attach too much heroism and bravery to their commemoration. The Ruhehain achieves this by acknowledging the need to remember those who were killed but suggesting that they are but one part of the community's history and purpose. The cemetery's official website mentions little of its history, though it does have a short piece on its original purpose from 1920: "For generations," the website says, "citizens remember those fallen in war on the National Day of Mourning. With the creation of the natural cemetery, though, this place became much more. Since 2008, even individual memorials have been held at the lectern."[1]

As an American war veteran now living in Germany, I have been fascinated by the comparison of our two memorial cultures. The Ruhehain's assertion that what was once a war memorial "became much more" by expanding to the public is at odds with the American veneration of military service above nearly everything else; the strict conditions required to be buried in Arlington National Cemetery or other, smaller veterans' cemeteries around the country. Americans tend to construct barriers—both physical and sentimental—between those who serve and those who don't, bombarding their citizens with so many images of war that they become easier to forget. The Ruhehain does the opposite, creating a space to mourn the fallen within an environmental setting whose intended appeal is to outlive any individual memories of those who are buried there. Yet the original war memorial, impossible to miss as you walk up a small hill into the shade of the oak trees, sticks out as a clearly man-made object in the otherwise natural woods, perhaps meant as a subliminal hint to visitors of mankind's role in the need for war memorials to begin with.

In his book *Sites of Memory, Sites of Mourning: The Great War in European Cultural History*, historian Jay Winter noted that memorials across Europe erected in the immediate aftermath of the war served multiple functions and could be found in and among the communities that built them:

> War memorials were places where people grieved, both individually and collectively . . . To find them one must simply look around. The still visible signs of this monument of collective bereavement are the objects, both useful and decorative, both mundane and sacred, placed in market squares, crossroads, churchyards, and on or near public buildings after 1914 . . . They have a life history, and like other monuments have both shed meanings and taken on new significance in subsequent years.[2]

In Germany, these sites of memory are everywhere, if easy to overlook. In most cases, local memorials blend into the town's scenery. If I ride on the left side of the bus on my route home from work, I'll just catch a glimpse of a stone pillar with an iron cross and the words *Zum Gedenken der Gefallen* ("In Memory of the Fallen") above the familiar year groupings, 1914-1918 and 1939-1945. A low stone wall covered with flowers and bushes separates the memorial and the sidewalk. If I blink and don't look in time, I'll only see the local bank. In Amberg, the town where I lived while stationed here with the Army, another easily missed memorial hangs under the shade of an old balcony attached to the town hall. There are 768 names listed next to a more elaborate inscription than most German war memorials: "To the everlasting memory of the heroes who died for their fatherland in the World War, 1914-1918, dedicated by the grateful city of Amberg." There are no names from the Second World War.

What is it about lists of names engraved into metal or stone that causes us to forge a more emotional connection with their loss? Perhaps we are more likely to see a part of ourselves in that dark, mystical space between the sets and rows of printed letters. We can imagine the person, the soul behind each name, the lives they lived and the deaths they met. The feeling is more powerful when confined to the local level, when the lists of names represent people who tilled the same soil, walked the same streets, went to the same church, and even lived in the same buildings as the town's residents a century later. Whether separated by days after the war or decades, the image of a local community tending to the memory of their fallen is, in some ways, more powerful than something larger and impersonal that one might find at a national memorial.

When I took my daughter for a walk to the Ruhehain shortly after moving to the nearby town, I couldn't help but read each of the names on the plaques to see if my family name, which has German ancestry, was among them. It wasn't. But while searching for information and history about other sites, I came across a digital archive of war memorials with a searchable database by name or town across the entire country. Curious, I looked up the memorial located in the town where I had lived previously, a site that I had somehow overlooked while living there. I found a Heinrich Eissler, born January 25, 1899, and killed October 18, 1918—just a few short weeks away from the armistice. I then put in a more general search for any Eisler or Eissler whose name appears on a memorial to the fallen from either World War.

There were dozens.

David F. Eisler is the author of *Writing Wars: Authorship and American War Fiction* (University of Iowa Press, 2022). In 2021 he completed a PhD in American studies at the Universität Heidelberg. Between 2007 and 2012 he served in the United States Army, earning the rank of captain, and completed overseas tours in Germany, Iraq, and Afghanistan. David received a BA in astrophysics from Cornell University in 2007 and an MA in international affairs from Columbia University in 2014. His academic writing, personal essays, and creative works have appeared in a variety of outlets, including the *New York Times*, *War on the Rocks*, the *Daily Beast*, *Collier's Magazine*, *Military Review*, *Drunken Boat*, and the anthology of short fiction, *The Road Ahead: Fiction from the Forever War* (Pegasus Books, 2017).

Notes

[1] "Naturfriedhof Reichartshausen," Link at: Startseite von Naturfriedhof Reichartshausen, accessed January 18, 2022, https://www.ruhehain.info/seite/517875/gedenken.html>. Translation by David Eisler.

[2] Jay Murray Winter, *Sites of Memory, Sites of Mourning: The Great War in European Cultural History* (Cambridge: Cambridge University Press, 2010), 79.

Demonizing the Enemy: "One in the Same"

ERNEST LUCAS MCCLEES, PhD, veteran

Destroy This Mad Brute -Enlist. Poster by Harry Hopps. World War I Posters Collection, Library of Congress Prints and Photographs Division, Washington DC, LC-DIG-ppmsca-55871.

With all conflicts, from WWI to Afghanistan, one question endures: When the United States goes to war, who is it going to war with? As a nation was created as a refuge for people from other nations around the world, how dangerous is labeling someone as the "other?"

As the 19th century morphed into the 20th century, the United States' population was growing rapidly, with many immigrants from eastern and southern Europe. These immigrants made the changes possible for the United States to expand its emerging industrial economy. The burgeoning industry and increasing ethnic minorities were both indispensable in shaping a nation on the cusp of financial and social power. The European minorities were diversifying the United States religiously and culturally, but they were also busy assuming their newfound national identity. Native languages and customs co-existed alongside the working language of English and modern American rituals. This functional climate succeeded, but as the United States turned toward war, so did the positive social advancements and social progress.

WWI propaganda posters such as "Beat the Hun" or "Destroy this Mad Brute" dehumanize the troops of Eastern Europe in a sardonic light. In tandem with various social programs, this propaganda created distances between new Americans. It not only harmed the progress of former Eastern Europeans and their families but of other immigrant groups as well. Lines of division and social injustice increased during a time when it would have been more productive to build cultural bridges for propelling universal assimilation or, at least fostering an understanding of the newer population who had become the backbone of the economy. Maybe it would have been possible to correct social unrest and problems by redirecting the propaganda during the war.

Contemporary America seems to mirror the problems immigrant groups faced during WWI. Disturbing parallels abound. Multinational immigrants and their children who contribute to the current economy are under fire with the growing threats of restrictive immigration laws. Hate crime is on the rise. Propaganda furthering a human divide with negative assumptions and imagery. Today, we find community watch parties, slogan posters, a printed recent cousin of yellow journalism, organized hate groups, and mass-hysteria xenophobia influenced by propaganda in every form of media ranging from print to live coverage. All seem to have an agenda, and currently, it looks as if unity is not a strong selling point.

Sadly, this is not a time to highlight differences, but similarities, and to encourage a public understanding of the process of enculturation, to learn from one another. In a world with a globalized economy, people live and work in multinational situations now more than ever before. The United States and its economy are dependent on these overlapping fabrics. During the Democratic National Convention in the summer of 2016, Khizir Khan challenged a presidential candidate by telling the story about his son who died fighting in Iraq as a Muslim American Army captain. While this example drew much attention at the time, it only started the current dialogue about cultural and religious divides. Since then, the dialogue has intensified, polarizing geographical, religious, and cultural differences. In the name of prosperity and true egalitarianism, it is time that nations with multinational backgrounds and identities win the internal war among themselves when they fight wars abroad. This "double war" was a gargantuan dilemma during WWI and remains the same today.

Dr. Luke McClees is a co-creator of the world's first Veteran Studies major. He also is the host for the podcast Veterans in Academics. Before Saint Leo University, he taught, presented, and published for a decade for the College of Education and Veteran Studies Program at Eastern Kentucky University. Dr. McClees served as a sergeant in the United States Marine Corps with the Third Battalion, Eighth Marine Regiment at Camp Lejeune, North Carolina. Dr. McClees served in both mission-oriented and rotational deployments. These deployments included a combat action deployment to help end the genocide during the Kosovo War.

The Enemy You Killed

RUTH EDGETT

It's April 17, 2013, and a magnificently blue sky arcs over the open fields of Northern France, where I've come to retrace my Canadian grandfather's journey as a young soldier in the Great War of 1914-1919.

The countryside sweeps wide in all directions here, a peaceful patchwork of tilled and greening fields in the warm spring light. It's hard to believe that all of this was once a churning battleground, riven with trenches and tunnels, writhing with soldiers and littered with dead and injured. It dawns on me now that, here in this region, the words "battlefield" and "cemetery" mean the same thing.

I'm standing in the *Duetscher Soldatenfriedhof* at Neuville-Saint-Vaast. Miles away across the plain I can just make out the towering white pylons of the Canadian monument at Vimy Ridge, the scene of Grandpa George Millar's proudest battle. Beneath my feet are buried 45,000 German soldiers: the 38,000 named ones lie four to a metal cross; the 7,000 unnamed ones, shoulder-to-shoulder beneath squat stone crosses in one long mass grave.

Less than a mile north and on the other side of the Arras-to-Souchez highway—once at the front line—is La Targette, where acres of white crosses mark the resting places of 11,000 French soldiers.

A couple of miles farther on is Cabaret Rouge, the burial ground for nearly 8,000 British Commonwealth soldiers, including Canadians. Their rounded headstones follow a familiar regimented pattern: arrow-straight row upon row. A mile or two farther off to the east is Vimy Ridge itself, where more Commonwealth dead lie in precisely-drawn burial grounds near the massive white monument that commemorates my country's loss of 60,000 young men to the long and grinding conflict misnamed The War to End All Wars.

In this place where lie many of the Germans my grandfather helped push back from Vimy Ridge, the grass is just as green, the trees just as vibrant as in the other cemeteries—but something is different. While the Allied burial grounds throw themselves open to passers-by and beckon from towering memorials and elaborate gates, this largest German cemetery in France languishes in relative obscurity. Sitting quietly behind berms and leafy trees, it's easy to miss from the road.

The grounds are tidy and mown as in all the other cemeteries—indeed, I can see people and machines busy at work just now. But this place is not as carefully manicured as the cemeteries of the British, or the Canadian, or the French. The grass is not quite so short-cropped and level, the lawn's edges less meticulous. Trees just coming into leaf and full, tall pines intersperse the long, straight rows of grave markers, and there is a sense that this place is allowed to simply exist—not necessarily as a destination for pilgrims and tourists, but as a thing-in-itself, here for its own sake.

This Duetscher Soldatenfriedhof was created under terms of the 1919 Treaty of Versailles, which allotted burial sites on French soil to all combatants of WWI. The German War Graves Commission, or VDK, administers this one, along with three others. It rests near a one-time warren of German tunnels and trenches dubbed "The Labyrinth" for their treachery and complicated layout. The town itself was obliterated by the fighting and reconstructed after the war's end.

It becomes obvious standing here that, while the Allied countries make much of the sacrifice and valor of their young men, their former enemy, humble in defeat, mourns quietly. But these soldiers had wives and children, mothers and fathers, brothers, and sisters at home, too. They called upon that same love of country, that same well of courage.

In *Battleground Europe: The Battle for Vimy Ridge–1917,*[1] authors Jack Sheldon and Nigel Cave reprint one German soldier's account of a particularly hard-fought engagement at Hill 145, the exact location on the ridge where the Canadian monument now stands.

The Battle began at dawn on Easter Monday, April 9, 1917, with a precisely-timed and hours-long creeping barrage of heavy artillery meant to beat the Germans back and provide cover for the initial advance of Canadian troops. But there were moments when that apocalyptic shelling let up. It was during one of these lulls that a company of Germans raised their heads above the parapets long enough to make out waves of Canadians advancing on foot.

"At long last!" wrote *Feldwebel* Paul Radschun. "Now there was going to be a battle with the same weapons, on even terms, shot for shot, throw for throw."[2]

Although the Canadians won the battle, the Germans held their fall-back line, and this is possibly what the sergeant was referring to as he concluded his hyperbolic narration: "It had defended its appointed place to the last drop of blood; worthy of its fathers; worthy of its parent formation the Prussian Guard; worthy of the heroic spirit of its beloved commander, who had always taught it to stand firm against the odds in all circumstances."[3]

The 100-hectare Vimy Memorial Park receives a million visitors a year. They come from around the world and all walks of life. Many are students on school trips from England or Commonwealth countries. These same tourists visit the cemeteries nearby at La Targette, or Cabaret Rouge—or Notre Dame de Lorette with the remains of 40,000 more French soldiers, some of whom died in earlier attempts to take back Vimy Ridge.

My grandmother Ruth kept a diary of the pilgrimage that she, Grandpa, and some 7,000 other Commonwealth veterans made to Vimy Ridge in 1936 for the unveiling of the Canadian monument.[4] In it, she wrote of meeting an elderly couple from Saskatchewan intending to visit the grave of their son whose heroism had earned him the Victoria Cross. She spoke to a young man on his way to the burial site of the father he never knew. But she seemed most affected by a widow whose husband had enlisted soon after they married. He would not live to see the son born a few months after he'd shipped out. Of this woman, Grandma wrote: "She had the privilege of visiting his grave twice and she said to me: 'As I stood there at his grave he seemed so near but...' and the tears filled her eyes and a break came in her voice as she added, 'so far away'."[5]

Isabelle, my French guide, says most tourists don't want to know about the dead German soldiers, even now. In fact, she's surprised I've agreed to have a look. Why would I not, I say, and ask how she feels about the soldiers beneath our feet, who invaded her country—whose sons and grandsons and nephews repeated that aggression a generation later, though that was a war these men knew nothing of.

She shrugs: "I think they were just doing their jobs."

But, why do so few Germans visit these last resting places of their countrymen, I ask. Why don't they feel the need to trace—like the Canadians and the British and the French—the historic trails of their ancestors? Isabelle once asked that of a German journalist, who responded that most likely shame that keeps them away, the shame of the second war bleeding backwards to stain the first.

A year after my visit here, Canadian journalist Matthew Fisher will write about exactly this; about how he "staked out" this very cemetery for three days hoping to speak to German visitors and learn their impressions about the centennial of World War I, which was commemorated all across France for the next five years.

"I utterly failed," he writes.

During his time at this cemetery, he meets a handful of visitors, mostly from Britain, some from Belgium and The Netherlands but none from Germany. Meanwhile, he says, thousands of tourists—including a bus load of German students—swarm the Vimy park to learn the Canadian version of how the first Great War was fought.

Fisher speaks to one British tourist, who is moved to tears by the enormity of the German losses made evident in this place.[6] Malcolm Tebett speculates about why no Germans have come: "I guess they feel there is nothing for them here."

All the same, this cemetery has a guest book; and, if either Tebett or Fisher opens it, they will find my name inside.

Some accounts of WWI raids and battles at Vimy Ridge read like play-by-plays for evenly-matched sports teams. Through them runs an undercurrent of respect, even admiration, for an enemy tenacious enough and clever enough to have held that strategic Ridge for three years despite repeated and costly attempts to dislodge them.

There are stories of soldiers making peaceful pacts across the front lines, and numerous ones about Christmas, like the night a handful of Germans emerged from their trench singing and celebrating, inviting the Canadians to join in. The Canadians declined but left their daytime enemies to mark the holiday in peace.

Authors Sheldon and Cave use the words of soldiers themselves to illustrate what one truce felt like. It was March 1, 1917, and the Canadians had been preparing for months to mount the battle for Vimy Ridge, a hotly-contested strategic high point that had been held by the Germans since 1914.

Even as the Canadians planned the big fight, they mounted random raids on the Germans hoping to wear them down and gather intelligence. Some raids were successful, but some were not. The overnight raid of March 1 proved particularly disastrous, and as daylight bloomed through the fog of a now-quiet battlefield, 687 Canadians lay dead and wounded.

Lieutenant Clifford Wells was picking his way through no-man's-land, searching for the wounded when he met a German stretcher bearer: "He offered to guide me to a number of *'verwundete Englander,'*" Wells wrote in a letter to his mother.[7] The lieutenant put together a stretcher party and collected twenty, but not without the help of five Germans and the loan of an additional stretcher, which was later returned.

Though Wells cautioned his mother not to repeat the story to anyone outside the family, he wrote of how much he admired the German soldiers, not only for their skill and endurance, but for their willingness to help carry Canadian soldiers to safety under threat of being sniped from the Canadian lines.

"They treated me with great respect, calling me *'Herr Leutnant,'*" Wells wrote. "We saluted before parting."[8]

That same day on another part of the battlefield, *Oberstleutnant* von Goerne, commander of German Reserve Infantry Regiment 261, set out in the fog with another officer to determine for himself whether reports of "enormous" enemy casualties were true: "Here, hard up against the barbed wire, we came across the dead Canadians, who were indeed very numerous," he wrote later.[9]

He and another officer continued into the center of no-man's-land to survey the overnight carnage. They managed to convince some Canadians, who initially attempted to shoot them, " . . . that they should cut out that stupid firing,"[10] and bring forward a superior officer. Word was passed up the chain of command and both sides agreed to call a truce until the dead and wounded could be taken back to their respective sides of the line.

While Germans and Canadians worked side by side, their officers stood together in one spot sharing German cigarettes. At 2 p.m. they shook hands and returned, with their men, to their respective positions. Soon afterwards a Canadian officer appeared in the German trench to thank them, and to offer reciprocation if the Germans should later find themselves in a similar situation.

"I must state that the Canadians, by their upright military bearing and their behaviour, made an outstanding impression on us," wrote von Goerne. "They could almost have been from the 261st!"[11]

The Canadians learned much from that doomed March raid. When the Battle of Vimy Ridge finally did take place on April 9, it was the Germans whose effort fell short—not tremendously short, but just short enough.

Historians often refer to WWI as a war of attrition. War diaries and other accounts of the Battle of Vimy Ridge provide fitting examples. Fights over small scraps of land were long and grinding affairs, and a victory won today might be taken away tomorrow. But for a few more reinforcements here, a little less ammunition there, the Battle of Vimy Ridge—even the entire war—might have ended much differently.

As I wrote these words in 2018, I thought about Grandpa and of all the barely-men who faced each other across those battlefields. And I think of these words from a Wilfred Owen poem displayed in a war museum not far from where those soldiers are buried:

"I am the enemy you killed, my friend."[12]

Ruth Edgett is the author of *A Watch in the Night: The Story of Pomquet Island's Last Lightkeeping Family* (Halifax: Nimbus Publishing, 2007) and of stories in other Nimbus collections. Her Maritime Canada-themed fiction and non-fiction appear in magazines, books and journals in Canada, the U.S., and the UK. Winner of the 2018 *CONSEQUENCE Magazine* "Women Writing War" fiction prize, Ruth is also a debut novelist. She was born and raised on Prince Edward Island, Canada, and received her higher education in both Canada and the U.S. Ruth Edgett lives and writes in Dundas, Ontario.

Notes

[1] Jack Sheldon and Nigel Cave, *The Battle for Vimy Ridge, 1917* (Barnsley: Pen & Sword Battleground, 2007).

[2] Sheldon and Cave. *The Battle,* 223.

[3] Sheldon and Cave, 223.

[4] After I returned home from France, that diary and my own pilgrimage to Vimy Ridge formed the basis of a short story called "Hill 145," winner of the 2018 CONSEQUENCE Magazine "Women Writing War" Fiction prize.

[5] Ruth E. Millar, "Ruth Reminisces," (1936).

[6] Matthew Fisher, "German Commemorations of First World War Are Muted," 2014, http://ww1.canada.com/after-the-war/vast-numbers-of-german-war-dead-lie-buried-in-france-scarcely-remembered-by-their-countrymen.

[7] Sheldon and Cave, *The Battle,* 53.

[8] Sheldon and Cave, 53-54.

[9] Sheldon and Cave, 126.

[10] Sheldon and Cave, 126.

[11] Sheldon and Cave, 128.

[12] "Strange Meeting by Wilfred Owen - Poems | Academy of American Poets," Poets.org (Academy of American Poets), accessed January 19, 2022, https://poets.org/poem/strange-meeting.

The 2014 Sainsbury's Christmas Truce Advertisement: Cinematic Memorial in Action

ANNA RINDFLEISCH

In 2019, following the centenary of the signing of the Armistice, the focus of the lasting legacy of the Great War shifted to a discussion about the ways Europeans exited the "War to End all Wars." The massive outpouring of social media postings and institutional centenary events from 2014-2018 suggests that the more than 100-year-old trauma attached to the iconic image of the Front Soldier was transmitted down generations, shaping our contemporary understanding of the Great War.

The BBC published a shocking video-short in 2015, "How Much Do Millennials Know About WWI?" The WWI knowledge of European millennials reflected the atmosphere of those who lived a hundred years ago; they knew it was fought mainly in Europe, it had been a horrific event, a lot of teenage boys were sent to fight, a lot of them were killed, and Germany lost. As to why the fighting started or what its ultimate purpose was, they had no words.

THE UNKNOWN SOLDIER AND LES MONUMENTS AUX MORTS

The mediated memory of the Great War has always emphasized its human and psychological cost by commemorating the loss of the war-dead. We can understand mediated memory as representations of the past transmitted through modern media that affect the construction of personal and/or collective memory. The BBC video, coupled with the public centenary ceremony that the Royal British Legion and 14-18 NOW put on in 2014 demonstrated just how present the traumatic memory of the Great War is today. To commemorate the very hour Britain entered the Great War in 1914, London was plunged into a dark silence on August 4, 2014 in the "Lights Out" ceremony. At 10 p.m., people flocked in the darkened streets to a candlelight vigil at Westminster Abbey. It ended at 11 p.m. with the temporary stamping out of the torch at the base of the Tomb of the Unknown Soldier. The ceremony included musical orchestration and readings of dead soldier writings broadcast over loudspeakers. On the 11th hour on November 11, 2018, a similar event commemorated the ending of the Great War. A two-minute silence was observed at 11a.m. to mimic the guns falling silent. Massive celebrations, commemoratory ceremonies, and social media selfies with people wearing poppy pins erupted around the world for the rest of the day.

At the crux of each event was the burial of a single unknown soldier. As such, this performance of memorializing war-dead was once again, a hundred years later, focused on a ritualized mourning practice implemented around the world. Laurence Van Ypersele in the chapter he contributed to A Companion to World War I entitled "Mourning and Memory, 1919-45," argues this figure of the unknown soldier is a potent focal point because its anonymity is and was able to encompass "the loss of an entire nation" which in turn "guaranteed equality for all heroes and allowed for the mourning of each individual."[1] This is the same process applied when interacting with les monuments aux morts (monuments to the dead), public war monuments that bear the names of individual fallen soldiers. The bereaved visit these monuments as many times as it takes for them to, in theory, complete their grieving.

Both to mourn the 1914 beginning of WWI and commemorate its 1918 end, centenary organizers who focused on unknown soldiers and lists of names in 2014 and 2018 were deploying the same tactics used over 100 years ago. These mourning rituals which emerged in Europe during the interwar period still seem to influence the way Western Europeans understand war commemorations today.

THE LANGUAGE OF SILENCE: AUDITORY COMMEMORATION AND PERFORMATIVE NON-SPEECH ACTS

WWI historian Jay Winter in his recent book *War Beyond Words: Languages of Remembrance from the Great War to the Present*, discusses auditory commemoration as key to remembrance. He defines auditory commemoration as "performative non-speech acts," which are "constitutive rather than descriptive," meaning "they speak rather than describe."[2] He provides the two-minute periods of silence that accompany Armistice Day ceremonies as examples. These performative non-speech acts and the silence that defines them appears reverberating throughout modern-day U.S. military ceremonies like the Last Roll Call ceremony, Missing Man formation, Twelve Gun Salute, and The Tomb of the Unknown Soldier's Guard at Arlington Cemetery. This type of silent mourning emerges at the "Lights Out" vigil at Westminster Abbey and the darkening of London at 10 p.m. on August 4th, 2014. These events suggest silence is a universal language that gives special non-verbal speech status to memory, perhaps giving expression to internal unspeakable trauma of war. Acts like the two-minute silences at military ceremonies are perhaps ways for people to tell interpersonal stories of their war experience with what Winter says are "beyond words."

THE 2014 SAINSBURY'S CHRISTMAS TRUCE ADVERTISEMENT: CINEMATIC MEMORIAL IN ACTION

NOTE: *As of January 31, 2022, the Sainsbury Christmas Truce 1914 ad discussed below can be viewed on YouTube.*[3]

Professor Mark Connelly from the University of Kent approached the topic of mediated memory within public consciousness at a talk he gave in December 2015 about the Sainsbury's 2014 Christmas advertisement, which tells the iconic story of the 1914 Christmas Truce. In the video, a soldier receives a chocolate bar from his love-interest back home. Emotion spurs him to venture out of the trench and into No Man's Land where he and a German soldier shake hands. The rest of the soldiers clamor out of the trenches and play a game of soccer before the rumbling sounds of gunshots sends them retreating to their sides.

The Sainsbury Christmas advert exemplifies cinematic memorial in action, thanks to modern-day technology and the fad of "Reaction Videos." Reaction videos are videos in which people react to events. In particular, videos showing the emotional reactions of people viewing television series episodes or film trailers are numerous and popular on video hosting services, particularly YouTube. The ad impacted the UK public visually as well as textually. If parsed, frame-by-frame, the advert becomes something of a micro-history on war memorials and its relationship to the modern-day use of war memory.

Darkness fills the first frame, then text appears on the screen and the light from the lantern brings men in uniform into focus. "Silent Night" begins to resound as the camera pans upward revealing the trenches. German and English overlap as the men from both sides sing a universal song of the Holiday Season. The viewers consume these emotionally charged images, the text, and the song—all are associations society has learned about the experience of a Great War trench soldier. The familiar sound of "Silent Night"—a song still popular at Christmas Time today—secures an emotional connection between the language of memory being portrayed through the symbolic imagery and the viewer.

The camera hones in on one soldier, who's received a package from a runner. He opens it slowly. Then he holds up a picture of a woman and the camera captures the slow upturn of his lips into a smile. Underneath the picture appears a letter and a chocolate bar. The soldier regards the chocolate bar fondly. His dirty fingernails and well-worn gloves correspond to the common stock-image of a Great War soldier, an image we all know, in a state of filth and despair. The appeal to this collective memory allows the viewer to validate the film as a truthful representation of a WWI British front soldier's appearance. The contrast of the chocolate bar's blue packaging, the dirty fingers of the soldier, and the nostalgic smile the soldier displays to the camera convey to the viewer that they are witnessing an emotional response to a gift his sweetheart has sent him.

The soldier seems to have come to a decision. He stands up and ascends a ladder. To the left of the screen a comrade shouts "Jim! Jim! No, don't do it!" Jim pokes his head out from the trench, hat held up as a symbol of peace, his face wrinkled in fear. Just a mile away, the Germans ready their guns until a single soldier shouts "Stop! He's not armed!"

Jim and the German soldier emerge from their trenches inching towards each other. Their respective battalions follow, and the two soldiers reach the middle of No Man's Land. "My name is Jim," he says. Silence, then the German soldiers respond, "My name is Otto." They clasp hands to an eruption of joyous music. The soldiers close ranks, remove hats and jackets as they shake hands. The viewers feel elated.

Jim shows Otto the picture of his sweetheart, whose holiday gift sparked the temporary truce. Otto remarks on her beauty. Within this frame the viewer enters into the private lives of two soldiers through this image of a woman.

Shouts and laughter pull the soldiers from their conversation. They dissolve into the mass of teenagers engaging in a game of soccer. They appear just like any other group of boys playing a game in an open field. The destitution of the war glimpsed at the beginning of the advert has fully disappeared.

A rumble of bombardments in the distance breaks up the laughter, reminding soldiers of the reality of war. Quickly the noise of war drowns out the laughter, smiles, and soft sounds of upbeat music. Officers from both sides look off into the distance, a steely look of resolve slips back into place. The smiles have all but vanished now, the two soldiers that started the truce shake one final time, and both sides disappear into their trenches once more.

As soldiers slump back into the earthen walls, Otto reaches into his pocket and, to his surprise, pulls out the candy bar Jim received from his sweetheart. He stares down and the viewers watch, transfixed, as the emotional climax reaches its apex.

In the last scene, Otto looks up into the distance, a small smile of shock spreading across his face. The camera pans toward the blue of the sky, as birds fly across the screen, and the final text flints onto the screen before the advert ends. It reads: "Christmas is for sharing." An unspoken reality of this advert rings loud; in 1914, these soldiers had to resume firing on the same men they'd just played soccer with. Despite the warm feelings from the football game and the shared chocolate, his historical realization also invokes the monstrous horror of the Great War and settles on the viewers in a way they may have never experienced. In the advert, the human aspect of the war was emphasized first, which made the historical reality more impactful.

Winter argues, "In 1914, war had a human face."[4] Indeed, it did. The Christmas Truce of 1914 is, in many ways, the emblem of Winter's quote, the very reaffirmation of the War that toppled the traditions of warfare and required the refashioning of mourning practices.

By 2018, the advert had received over 20 million views on YouTube with over 12,000 comments, the vast majority discussing emotional and visceral reactions. One commenter, Sabrina Umstead, wrote: "My 7th grade social studies teacher showed our class this while learning about World War 1. This is forever my favorite commercial not only because of the story behind it but the memory attached."[5] The advert's use of the Christmas Truce influenced a teacher so much that he worked it into his lesson plan for the Great War. Umstead's quote brings together the message of the ad's end, "Christmas is for sharing," and the traumatic memory of the War. In the "TV Reactions to Sainsbury's Advert" video that Ed Ward released on YouTube December 6, 2014 similar visual responses of Londoners in tears reiterate these commenters' sentiments.

Advertisements function differently than films, novels, poems, or plays. Ads are marketing tools aimed at relaying a specific message. They must sell a product or ideal. In the case of the Sainsbury advert, the product was the chocolate bar. In the YouTube description attached to the Sainsbury advert, the company wrote: "The chocolate bar featured in the ad is on sale now at Sainsbury's. All profits (50p per bar) will go to The Royal British Legion and will benefit our armed forces and their families, past and present."[6] Coupled with the emotionally jarring experience of the advert and the marketing push, on December 17th, 2014 Sainsbury's Twitter feed released a picture of the chocolate bar with the caption: "We're on track to raise £500,000 for The Royal British Legion through our chocolate bar sales! #ChristmasIsForSharing."[7]

Sainsbury's ad was released to YouTube on November 12th, 2014 and, within 38 days they turned a massive profit of approximately $657,980. Two days later, Sainsbury had doubled their original production of the bar and responded to a commenter on Twitter who hadn't been able to find it in any of the stores near him. Sainsbury replied: "The demand for these was greater than we ever expected. Really sorry to hear you weren't able to get one."[8] At its peak, Sainsbury was selling 5,000 chocolate bars an hour and had over 10 million views within the first week of releasing the ad.

With massive popularity came an equally daunting backlash. Neil Kelly was quoted by Rebecca Perring, a writer for *Express, Home of the Daily,* and *Sunday Press*: "It's a lovely story from history but I find it upsetting that they've used the First World War as a vehicle to promote a supermarket. The sentiment behind it, supporting the RBL, is sound, but there's something that doesn't sit right with the use of the war."[9] This response was just one of the thousands of people who saw the advert as a sacrifice of the Great War's memory to consumerism. The ad's immediate popularity and the massive response Sainsbury received proves the Great War vividness in modern-day thought.

The BCC video-short released in 2015, the Royal British Legion's "Lights Out" ceremony held in 2014, the widely popular Sainsbury Christmas advert, and the outpouring of events and social media postings commemorating the centenary demonstrate the success of the Myth of the War Experience and the failure of the mourning process in the Interwar Period. The Myth of the War Experience disseminated narratives of national purpose through the figure of the sacred Front Soldier to the public in an attempt at mass consolation and absolution of war guilt. The foundations of the Myth of the War Experience after the war built the National Public Grief narratives, which used the war-dead to address the universal need of the individuals' absolution of guilt. However, in allowing the bereaved to mourn collectively, the National Public Grief narratives might have removed their ability to grieve at a private level; thus, we see the resulting disruption to the mourning process. Perhaps this explains the intense, unexpected personal reactions to the Sainsbury's ad today.

If I informally postulate as to why we find these traces in modern-day thought, I come to multiple reasons. The rapid turn around from the end of WWI to the beginning of the Second World War limited the time for the bereaved to fully conduct the mourning process. The unprecedented atrocities of WWII perhaps overshadowed the lingering grief of its predecessor. A newly scarred community of the bereaved faced the trauma of a post-Hiroshima and Auschwitz world with a dismantled language of mourning. Both bound to language and betrayed by language inadequacies, the bereaved of WWII attempted to mourn in ways unseen before. Consequently, the bereaved of the Great War transmitted their unresolved, unspoken individual trauma into the forthcoming generations. Thus, there is the modern-day notion of the Great War as a monstrous affair which stole the lives of nearly 10 million teenagers as the BBC video-short "How Much Do Millennials Know About WWI?" revealed. Deeply impactful, the widespread popularity of the Sainsbury advert and the outpouring of sorrowful remembrance postings that accompanied November 11, 2018, unveils the strong emotional need within individuals to lament. The bereaved lack of closure during the interwar period may have endured into our contemporary society.

Anna Rindfleisch is a PhD candidate in English Research at King's College London, UK. Her research interests focus on interwar period mourning literature, gender studies, remains repatriation policies, and the transmission of war trauma into the performing arts. She has published on interwar period bereavement, gendered mourning rites, and the ramifications of war on society. She works for the U.S. Military's Department of Defense as a historian at the Department of POW/MIA Accounting Agency (DPAA) researching the last verifiable locations of the 7,774 unaccounted-for U.S. servicemembers who died on the Korean peninsula during the Korean War.

Notes

[1] "1914 | Sainsbury's Ad | Christmas 2014," posted by Sainsbury's, November 12, 2014, video, https://www.youtube.com/watch?v=NWF2JBb1bvM.

[2] J. M. Winter, "Shell Shock, Silence, and Memories of War," in *War beyond Words: Languages of Remembrance from the Great War to the Present* (Cambridge: Cambridge University Press, 2018), 172-201.

[3] "1914 | Sainsbury's Ad | Christmas 2014," comments.

[4] "1914 | Sainsbury's Ad | Christmas 2014," video description.

[5] Sainsbury's (@sainsburys) "We're on track to raise £500,000 for The Royal British Legion through our chocolate bar sales! #ChristmasIsForSharing," Twitter, December 17, 2014, 7:16 p.m., https://twitter.com/sainsburys/status/545281496756809729.

[6] Sainsbury's (@sainsburys) "The demand for these was greater than we ever expected. Really sorry to hear you weren't able to get one," Twitter, December 8, 2014, 10:57 p.m., https://twitter.com/sainsburys/status/545699305546604544.

[7] Neil Kelly quoted in Rebecca Perring "Tasteless! Sainsbury's face backlash for using First World War imagery in Christmas ad," *Express,* November 14, 2014.

Remembering and Forgetting: Some Photographs from a Small Corner of the Great War*

MARK FACKNITZ

1

He was my grandfather, my opa. Albert Karl Gustave Facknitz, born 12 September 1890, died December 1963. In October 1909 he began training in a marine unit, the naval infantry, joining the third Seebataillon, first Company; he was posted to Tsingtau (Qingdao) by February of the following year. He served for over a decade, including an eight-month leave and redeployment to Cuxhaven in 1913. During that time, he met Martha Zorn, one day my oma. He returned to China, served as consular guard in Tientsin, then returned to Tsingtau for its fall and was captured by the Japanese on 6 November 1914, the day before official surrender. After five years as a prisoner of war in Japan, Albert was repatriated to a Germany beset with the turmoil and shortages of the postwar period. He married Martha in November 1920. The photograph, taken by Japanese photographer Y. Kobayashi in Tsingtau, was printed on heavy stock and sent back to Germany, he sent to his sweetheart in Pomerania, inscribed, "To remind you of me, your friend Albert."

My grandfather was a private soldier, later a corporal, participating in the maintenance of Tsingtau, Germany's most ambitious project as a late-comer to imperialism. As early as 1871, the German geographer, geologist, and world traveler Baron Ferdinand von Richtofen (uncle of the Red Baron), ended his overview of the wealth and potential of Shantung Province with this exhortation: "Although the rise of the Empire of China is likely to antagonize European interests in intellectual, material and industrial respects, still this rise will become urgent from the pressure of necessity, and taking this fact into consideration the various foreign powers will each have to secure for themselves the greatest possible advantage in the approaching era of China's renascence."[1] Sun-Yat-Sen, after a visit to Tsingtau with its spacious streets, running water and sewers, modern hospital and docks and rail terminus, offered the opinion that Tsingtau was a model for Chinese cities of the future.

These are some of the photographs he brought back with him.

2

Among solid and stately administrative buildings, this cenotaph commemorates Germans, particularly soldiers, who died because of the German colonial experiment. That expedition of cultural and economic conquest began with two German Catholic missionaries, Franz Xaver Nies and Richard Henle who would long inform the mythos of German expansion in China. They were killed by a gang, presumed to be from the Big Swords Society, 1 November 1897, in Zhiang Jia village in western Shantung. Nies and Henle, on All Soul's Day, were visiting fellow priest Georg Stenz. Shortly after retiring for the night the mission compound was entered by killers. Nies and Henle were stabbed repeatedly. Within two weeks of their death, Germany landed an occupying force at Kiaozhou Bay, then compelled the removal of a regional governor, the building of three new Catholic churches, and a reparation of 3,000 taels. Within a few years, the singularity of the Juye incident, so-called, was blurred by the broader and more coherent Yihequan Movement, the Boxer Uprising, of 1899-1901. In the end, Germany exacted the Kiautschou Bay Concession, a ninety-nine-year lease of over two hundred square miles, which, while still belonging to China, were under German sovereignty. The agreement permitted construction of harbor facilities; a railroad connecting Shantung to Manchuria, and to the Tran-Siberian, and eventually Berlin; coal mining; and the planting of a modern German town.

3

The Boxer Uprising, cause and pretext for the European consolidation of power in China, also provided great and lasting rhetorical momentum. This photograph—"Execution of the Leaders of . . ."—dates from 1899-1901, ten years before Albert Facknitz's arrival. It shows Qing onlookers, decapitated Boxers, and Japanese officers and executioners. Like several of the photographs, this one was probably purchased, mounted on heavy paper, and titled in a careful hand in old-style German script. Two heads are missing from this copy. Who tore the photograph and why I do not know. I do know that my father as a boy would sometimes sneak into the cupboard and get out the pictures his father brought back from China. This one, he said, riveted him in its grisliness.

The presence of Chinese and Japanese soldiers in the execution photograph remind us of the tensions in the historical moment, particularly in China where the ambiguities of class, language, region, and decadence had not been sorted and simplified as they had been in the nationalizing fervor of Meiji Japan or, for that matter, Bismarck's Germany. Japan was among the eight nations which expanded their influence in China, especially Shantung and Manchuria, at the time of the Boxer Uprising. Such was the civil and military incoherence among the Chinese that under what circumstances the Chinese soldiers participated in the execution would be impossible to say with certainty.

4

Similarly enigmatic is this photograph of a European nurse—she wears the Red Cross medallion at her collar—flanked by women on either side. At first glance, the image seems simply to record the presence of western women in the colonial drama of the early twentieth century. Who she is, why she is in China, and exactly when the picture was taken and by whom, are details that I cannot recuperate. This recalls the conundrum of women's history; we know they were present in equal measure to men, but their documentation is sparse by comparison. So, in the absence of contextualizing narrative, frustrated in our desire to know who this young woman is with the modest smile and the good posture, we risk missing some startling details if we accept that her anonymity makes her unreadable. But as text, this photograph is remarkably rich. All the women to her right and the first two to her left are Chinese. The four at the right of the photograph, her left, are Japanese. The Chinese women are younger than the Japanese, not as well dressed, and while the Chinese look at the camera, none of the Japanese do. In other words, a complex cultural and circumstantial dynamic is at work here. That much we can know. What put these thirteen people in the same place and the same time?

5

This photograph my father once said he thought to be the gate to the old city of Tsingtau. It may have been anywhere, actually, and probably not Tsingtau, as the town was little more than a fishing village when the Germans arrived and began developing a port city, a railroad terminus, and an administrative and military center. By 1910, in China, Tsingtau was second only to Hong Kong in commercial traffic, amenities, cultural diversity, and strategic naval importance. It had massive and domineering new buildings in the Teutonic Imperial manner, but no significant remnants of Chinese antiquity. Still, notably in this photograph, the field of view divides between the awnings and shaded figures on the left (perhaps merchants?) and the waiting rickshaw drivers in the open light on the right. At midpoint, middle distance, there are three men. Two with umbrellas, two in European dress, talking and looking toward the camera.

Some others, equally evocative:

6

A small grouping of houses, presumably on the outskirts of Tsingtau. Schatzekon: sweetheart.

7

Sonnenuntergang: Sunset. One of the most poorly preserved of the photographs pasted to heavy stock.

8

Bambushain: bamboo grove. The condition of this picture is even worse, more blurry and bleached than the sunset. The adventure into China, while predictably exotic, seems also to have been sentimental, superficially exotic but archetypally European. A tidy hamlet, a sunset, and a grove mark the ambivalence between alienation and expropriation. What does it mean to fetishize the exotic even while assigning it familiar qualities characterizes several other pictures, still among those that appear to have been purchased and then mounted? Germany's project of building a modern commercial center and port city occurred while simultaneously colonizing West Africa and Samoa. Since it is not at all clear that my grandfather stopped in Samoa, it seems likely that his several pictures from there were part of a trade among other Germans.

9

In this, Weiberans Samoa, Samoan women, the woman with the elaborate headdress in the middle apparently bore a resemblance to my grandmother, or so thought my father and his older brother.

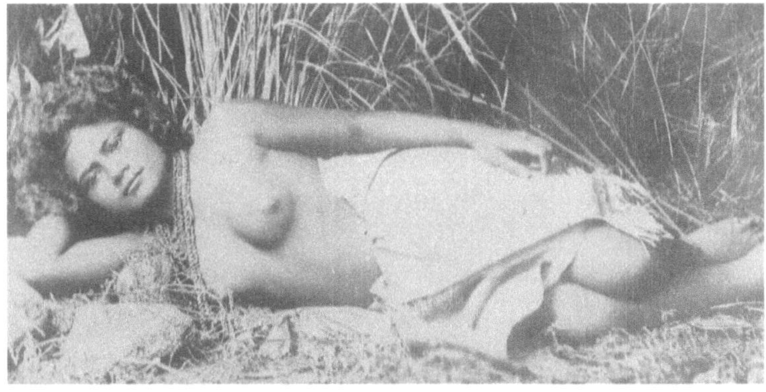

10

The appeal of this one to a soldier bivouacked among men in a foreign garrison is obvious. After release from prison in Japan, more than a hundred of my grandfather's cohort decided to remain in Japan. Many preferred to emigrate to Samoa.

11

Samoan princess? With headdress and knife, a bemused expression; it is not entirely clear that she knows what she is being asked to represent.

Other photographs in the collection are, it appears, one-offs, certainly not commercially produced, and some of these include my grandfather or members of his company.

12

The Germania Brewery, opened in 1904, provided Pilsner beer to the German soldiers and citizens of Tsingtau. In this picture the Chinese man with the palm fan, is not entirely sure of what he is doing or why he is in the middle. Today, the Tsingtao Brewery maintains a sizeable presence in the domestic Chinese market and makes one of China's principal export beers.

13

Sergeant Wilhelm Schmidt's squad. Everything suggests that the German troops in Tsingtau were exemplary in matters of morale and unit cohesion—disciplined, vigorous, and dauntless, they were young men who like my grandfather had eluded the near invincible determinants of nineteenth century poverty and were now beneficiaries of fresh air, camaraderie, ample food, dry bunks, and steady pay. In finding respect for country and fellow soldiers, they found self-respect. Opa was a sorrowful man; he was a man of unsurpassed dignity.

14

"The Fun Bunch," loosely put. The young Chinese male in the lower right seems to have been tugged abruptly toward the soldier to his left at the moment the photograph was taken. Higher up, more beer, and one additional Chinese person. The environment is of rocks and trees; nothing to remark an ambient culture, for the master plan of Tsingtau made China largely invisible behind the robust and vast Bismarck Barracks and spacious Catholic and Lutheran churches.

15

A picturesque and madcap set of poses at the seaside. The blurry single Chinese, bottom center, suggests that the photographer's command to hold still was sometimes lost in translation. A special limitation of the photograph as historical evidence is precisely the hegemonic blurring of the cultural topography, so novel and obvious are the contours of the physical landscape. Photographs are deceptive precisely because of their immediacy, their stability.

16

Spreading out over rocks—here at the seaside—seems to have been a motif. My grandfather is just above the penciled arrow.

With the assistance of the British, the Japanese began a naval blockade of Tsingtao at the end of August 1914, a few weeks after the declaration of war in Europe. In the first week of September Japan began to land the 18th Division of Infantry at Lungkow and Laoshan Bay. The Germans, outnumbered by more than six to one, prepared to defend Tsingtau. At this moment my grandfather, disguised in civilian dress, made his way by train back to Tsingtau from Tientsin, where he had been attached to a diplomatic guard unit. Well-provisioned, the Germans spent September and October waiting, establishing trenches, lines, redoubts, and artillery batteries. In 1915, American journalist Jefferson Jones wrote of Tsingtau that "Germany has . . . been forced to surrender all this magnificent work," and "the Tsingtau of 1914 will clearly stand out in the memory of its visitors and Far Easterners as the finest, the prettiest, most modern and sanitary city in the Orient."[2] Post-war, still believing in the promise of the city, several dozen of my grandfather's fellow POWs came back to Tsingtau to settle.

The photographs from this moment make plain the severity of the landscape as well as the remarkable similarity of German defenses in China to the trenches of the Western Front.

17

Machine gunners take up a position in the rocks and wait.

18

Front line troops before the siege. A defensive position that is downhill of the advancing infantry is not ideal, though this may have been an opportunistic position, either for the camera or because the soldiers could use agricultural terracing as breastworks.

19

From what appears to be a much more tenable position, a German lieutenant surveys the terrain.

20

Troops in a defensive drill.

Once combat began in earnest, October 31, 1914, the Japanese advanced steadily, and while the German garrison had ample food and medical provisions, they soon exhausted their artillery shells and were compelled to surrender. My grandfather's right cheek was grazed by a bullet; at one point debris from a shell burst landed so heavily on him that he asked a comrade if his legs were gone. In fact, they were merely bruised.

German casualties included two hundred dead and five hundred wounded compared to more than seven hundred Japanese and British dead, a ratio consistent with casualty comparisons between defensive and offensive troops in the European theater. Most of the German Imperial ships had previously left for Europe; however, a remaining cruiser, torpedo boat, and four gunboats were scuttled by their crews before the German surrender on November 7. All the men of the Third German Sea Battalion were taken prisoner and shipped to Japan. In Europe the sea battalions were put into the lines in Flanders and inflated to the size of divisions, eventually two marine divisions known as the Marine-Korps-Flandern, totaling seventy thousand troops. Had he been in Europe, rather than a POW camp, my grandfather likely would have been exposed to action at Yprès in 1915, or the Somme the next year, or in Flanders the year after that, and, on the slim chance he would still be alive, in the final desperate offensive of early 1918, he might have fought in the Kaiserschlacht. Instead, he experienced relatively good treatment in Japan, first for almost a year at Asakusa (now in central Toyko), and for the rest of the war and all of 1919 at a camp thirty kilometers away in Narashino.

21

The gatehouse of the Narashino prisoner of war camp. Compared to Asakusa—converted crowded and outdated Japanese barracks—Narashino, a large camp, purpose built, housed nearly a thousand, with sports facilities, camp library, and orchestra and chorus. My grandfather belonged to the Tischler-Innung Narashino, a cabinet maker's guild. In addition to numerous chess sets, which he gave to his captors and his comrades, he made a dozen violins. The boundaries of the camp were remarkably fluid; the prisoners worked with local farmers and gardeners, sharing ideas from German horticultural practices. (My grandfather treasured his rose bushes in later life.) Ordinary soldiers received half a yen a day which allowed them to purchase tobacco and soap.

22

The washhouse at Narashino, for bodies and clothing. Except for the influenza in 1918, staying healthy was not a particular challenge.

23

In large measure, the generous Japanese treatment of their prisoners derived from the captives' limited number and from the fact that in modernizing their military in the Meiji period, the Japanese had attended to Prussian successes, constitutional and military. They respected my grandfather's cohort as soldiers. Several pictures, such as this large bell outside a Shinto shrine, suggest his admiration for Japan and its culture. Once in America, to which he escaped from the hyper-inflation in 1926, bringing with him his wife and my oldest uncle and my father, the rise of Hitler would present him with a devastating sense that Germany's ultimate defeat was self-inflicted. The Japanese attack on Pearl Harbor deepened his grief, for standards of order and decency which he had endeavored to make his own would be lost in such bigotry and violence. From the post-World War II perspective—after the Bataan Death March, after Auschwitz—my grandfather's love of honor and organization, his trust in authority, seem remarkably antiquated, naïve. And yet.

24

In this photograph, he crouches in the middle of the second row. It could be a very ordinary picture of a company of soldiers in smart uniforms, all looking fit and healthy, assembled for an official photograph, were it not that the men in this photograph are all prisoners of war, inmates of the Narashino camp, who apparently had their dress uniforms in their kit when they left Tsingtau. They have assembled outside the boundaries of the prison to have their picture taken in front of a splendid Shinto temple. Yet, the historical momentum that culminated in the fall of Tsingtau was the same as that which took Rupert Brooke almost to Gallipoli, or led Remarque's Paul Baumer to the moment when, the last of his platoon, he stretched his hand toward a butterfly, or Wilfred Owen to the Sambre-Oise Canal where he died exactly a week before the Armistice. In our Eurocentric reading of the catastrophe, we ignore that elsewhere the catastrophe may read differently. In my family, it comes down to the simple fact that my grandfather had the great good fortune to be taken prisoner by the Japanese halfway around the world at the moment of the First Battle of Yprès in which eight thousand Germans would die, ten thousand would go missing, and nearly thirty thousand would be wounded.

25

With the author, Centerline, Michigan, in 1950. Opa's mother had died when he was a child; his father was remote and his stepmother a martinet. He was sent from the family's rough rural cottage in Pomerania to an apprenticeship with a carpenter at an early age. I remember him smelling of snuff, a quiet and melancholy man, an old soldier who was extraordinarily tender to his grandchildren.

*All photos from author's private collection are reprinted with permission by the author.

Mark Facknitz is Emeritus Roop Distinguished Professor of English from James Madison University. In recent years he has divided his research between the Great War and Willa Cather. His essay "Kitsch, Commemoration, and Mourning in the Aftermath of the Great War" appears in Jonathan Vance's *The Great War: From Memory to History.* He has published on Ivor Gurney's shellshock in *The Journal of the Ivor Gurney Society,* on war cemeteries and the margins of memory in *Bridges,* on Luytens and Thiepval as paradigms of commemoration in *Crossings,* and on gardens as surrogates for soldiers' remembered selves in *a/b Autobiography Studies* Facknitz's interest in the Great War largely depends on a German grandfather, prisoner of war in Japan 1914-1919; an American grandfather, an engineer in the AEF; and a great uncle who died for Canada. He holds an MFA from the University of Iowa and a PhD from the University of New Mexico.

Notes

[1] Ferdinand von Richtofen, "Notes on the Province of Shantung," in Robert Coventry Forsyth, ed. *Shantung: the Sacred Province of China* (Shanghai: Christian Literature Society, 1912).

[2] Jefferson Jones, *The Fall of Tsingtau, with a Study of Japan's Ambitions in China* (Boston: Houghton Mifflin, 1915) 153.

Beyond Friend or Foe.
World War I American Immigrant
Poetry: A Digital Humanities Project

LORIE A. VANCHENA, PhD

WWI poetry written and published by German immigrants in the United States reflects a range of perspectives on the global conflict, from strong support for Germans fighting against the Allied Powers to loyalty to the migrants' new homeland. This poetry, written in both German and English, appeared in the German-American periodic press and in anthologies. It sheds light on different aspects of the immigrant experience, including ways poets sought to define their ethnic identity at a time when public perception of Germans was changing dramatically, as evidenced—to give just one example—by the government order in 1917 that non-naturalized males aged fourteen and older register with the authorities as "enemy aliens."

Die-Wacht-Am- Rhein (The Watch on the Rhine). U.S. Food Administration, Educational Division, Advertising Section. Wikimedia Commons. Accessed November 6, 2021

One way German immigrant poets conveyed pro-German sentiment was to draw on nineteenth-century patriotic poems. They took up German cultural material such as Max Schneckenburger's "Die Wacht am Rhein" (The Watch on the Rhine) and refashioned it to address new historical circumstances. Written in 1840 in response to the Rhine Crisis, France's threat to reclaim territory west of the Rhine, Schneckenburger's poem suggests a German national consciousness at a time when a unified German nation did not yet exist. The patriotic text, with its refrain "Lieb Vaterland, magst ruhig sein / Fest steht und treu die Wacht, die Wacht am Rhein" (Dear Fatherland, no fear be thine / Firm and true stands the watch, the watch on the Rhine),[1] surged in popularity during the Franco-Prussian War 1870-1871 and again during the First World War.

Numerous poems in our collection allude to "Die Wacht am Rhein." Ludwig Lewisohn's poem from 1915, for example, which depicts Germans defending their Fatherland's western border, borrows the title "Die Wacht am Rhein" and cites part of the refrain to evoke recurring German-French hostilities (first two stanzas):

> Grim, gaunt and stern, in shimmering grey,
> The German lines advance,
> A thousand comrades' blood to-day
> Has drenched the fields of France.
> And though from barren haversack
> The last dry crust be gone,
> They wait upon their tireless track
> The cannonade of dawn.
>
> And from ten thousand throats ablaze
> With bitter burning drouth,
> The thunder of their hymn they raise
> And hurl it to the South,
> Unbroken stands the endless line,
> Unshaken heart and hand,
> "Lieb Vaterland, magst ruhig sein!"
> For firm your armies stand.[2]

Germania on Watch on the Rhine. Painting by Lorenz Clausen, 1860. (Wikime-dia Commons), accessed November 6, 2021

German immigrants reading the periodical in which this poem appeared, George Sylvester Viereck's pro-German English-language *The Fatherland,* were expected to recognize the German title and refrain. Lewisohn's references to the earlier "Die Wacht am Rhein" also urged readers to recall the Prussian victory against France in 1871 when forming their views of fighting on the Western Front. Although written for an English-reading audience, the poem reaffirms immigrants' German ethnic identity.

The majority of the 300 poems in our collection were written or published by Germans living in the U.S. Fully annotated versions of the poems, with scans of the original texts, are published on the project website: http://ittc.ku.edu/wwipoetry_project/poems/.

Dr. Lorie Vanchena, associate professor of German Studies at the University of Kansas, directs the KU Max Kade Center for German-American Studies. Her current research focuses on German immigrant culture in the United States in the early 20th century. She directs the WWI American Immigrant Poetry digital humanities project, a scholarly digital archive of poems written and published by German Americans during the First World War. She has published *Anton in America: A Novel from German-American Life*, an annotated English translation of Reinhold Solger's novel *Anton in Amerika* (1862). She teaches courses on the German-American experience and contemporary German-speaking Europe.

Notes

[1] Max Schneckenburger, "Die Wacht am Rhein." In *Deutsche Lieder*. J. B. Metzler, 1870, 19-20. *HathiTrust*. https://hdl.handle.net/2027/uc1.b3148536?urlappend=%3Bseq= 27%3Bowner-id=9007199272180709-31.

[2] Lewisohn, Ludwig. "Die Wacht am Rhein." *The Fatherland* 2, no. 18 (June 9, 1915): 10.

part seven

BROKERING A PEACE

When the War Didn't End:
"Polar Bears" in Russia and the Red Summer

ROB BOKKON

The stench of chlorine gas blew away into the wind. The clatter and slap of tank treads, the angry wasp hum of airplanes, the cries of the dying and the wounded all faded under a blanket of dull silence. Where before a pall of smoke had hidden the miles and miles of bloody mud and corpses and barbed wire, all came now into sharp relief as the great guns, at last, stopped their terrible song.

Four years of agony and boredom, heroism and stupidity, sacrifice and madness: all whisked away in the flourish of a pen.

Every WWI aficionado knows the date and the hour. 11:00 a.m., Paris time, November 11, 1918. The Armistice, and the end of the Great War.

And such an end: the Kaiser deposed; the German Empire defeated and humiliated. Austria-Hungary in tatters.

France safe and secure. Belgium avenged. Britannia triumphant, thanks to the help of her former colonies: young America, magnanimous in victory, ready to take her place on the world stage. Of course, there was the little matter of that unpleasant business in Russia, but it would soon sort itself out—for now, the war was over. The world was free from tyranny. Safe for democracy, in the words of President Woodrow Wilson.

This is the legend. This is the myth we have told ourselves and let ourselves hear. This is how we end the story, since stories must have a good solid boom at the end, if only to let us know when to applaud. Much like the memorable date and time of the Armistice, 11:00 a.m., on 11/11/18, this narrative is tidy. Precise. Simple.

The truth, as always, is not tidy nor precise and it is never simple.

In the words of Dr. Patti Hagler Minter: "Nothing in history is tidy. When you're given a simple answer to a complex question, look closer. Look at who's telling you the answer and why they want you to believe it."[1]

On the global scale, tragic stories abound about how the war didn't end on November 11, 1918. The many heartbreaking tales of the last few brave soldiers killed in the hours after the Armistice before word had reached every corner of this far-flung and truly global war. The brutal German campaigns in Africa, where General von Lettow-Vorbeck continued fighting for two weeks after the peace. The French occupation of the Rhineland and the racial unrest caused when German women "betrayed their race" by falling in love with, and marrying, French-Algerian soldiers.

THE U.S. "POLAR BEARS" IN RUSSIA

What is less well known is the involvement of American troops fighting the Bolsheviks in the nascent Soviet Union, and the lynching of African American veterans in the South, often by their own brothers in arms.

These stories are just as much a part of the Great War narrative as the tales of heroism at the Somme, or the resolve of the Belgians, or the grim determination of Britain's Gurkhas and Sepoys fighting for an empire that afforded them neither citizenship nor rights. It is our duty as historians to tell the truth inasmuch we can find that truth from the sources available to us.

Except for the alliance of WWII, eight decades of hostility existed between the United States of America and the Union of Soviet Socialist Republics. Yet almost all the fighting occurred in the form of proxy wars: Vietnam, Korea, the Malay States Emergency, and civil wars in South America. Though hundreds of thousands died serving the agendas of the two greatest world powers of the day, only rarely did U.S. and Soviet troops encounter one another, much less experience actual combat.

This hostility did not arise in a vacuum. Over the course of many months in 1918 through 1920, United States troops fought against Russians on Russian soil itself. This was the only time the two powers would ever meet in battle on home territory—and despite the decades-long global implications of this conflict, the story is virtually unknown.

Prior to the fall of the Czar, President Wilson arranged for over a billion dollars (in 1918 figures) of American munitions, rolling railroad stock, and military supplies to be delivered to Russia to aid in the Russian Empire's war efforts on the faltering Eastern Front. The supplies arrived in Vladivostok not long before the overthrow of Nicholas II in February of 1917. During the initial rule of the Provisional Government, Wilson was content to leave the supplies where they were, but the rise of the Bolsheviks in the October Revolution and the execution of the Imperial Family in July of 1918 changed the situation entirely, causing panic in the American government. Anti-communist sentiment, already beginning to build in the United States owing to sensationalist media stories of the Russian Revolution and the increase of strike activity among American workers, now reached a fevered pitch.

At the same time, the misadventures of the Czechoslovak Legion (CL) in Russia came to the attention of the Allies. A volunteer force, composed of rebellious subjects of the Austro-Hungarian empire who fought for the Allies, the Czechoslovak Legion's goal was to earn the liberation of their homelands from the Hapsburgs by war's end. Over 100,000 strong at its peak, the CL served as a unit of the Imperial Russian Army and racked up a string of victories to its credit but following the Treaty of Brest-Litovsk and the rise of the Bolsheviks, the Legion's commanders decided they would be better used on the Western Front. With most of Russia's ports under blockade, they attempted to travel to Vladivostok where transport ships would carry them to France.

The Bolsheviks, suspicious of a large armed force traveling through what was at that time largely unsecured territory, insisted on the surrender of most of the Legion's weaponry prior to granting safe passage. The Czechoslovak Legion reluctantly conceded to these terms in what became known as the Pensk Agreement. Despite this promise of safe-conduct, the CL took almost two months to cross Russia on the available railways. Their progress was hindered by the interference of local soviets, which often forced them to renegotiate the terms of their safe-conduct, and by official interference from Moscow.

The Bolshevik government, having fought the CL multiple times since the October Revolution, maintained a healthy skepticism of the CL's motives. And the Bolsheviks were right to do so: After their failed evacuation, the Czechoslovak Legion would remain in Russia for months, siding with the pro-Czarist White Russians in fomenting counter-revolutionary terror.

Despite the hardships of governmental interference and dilapidated railways, the CL eventually arrived in Eastern Siberia, with some units reaching their destination by May of 1918. By the time the bulk of their troops arrived, the strategic situation was dire: the Legion wound up strung out over many miles of the Trans-Siberian Railroad, harassed by partisans and Cossacks, and low on supplies. A clash with Magyar POWs at Chelyabinsk station finally caused the People's Commissar for War, Leon Trotsky, to decide that the government had had enough. He ordered the immediate and total disarmament and arrest of the CL, which responded by entering into hostilities with the Bolsheviks and seizing many miles of the Trans-Siberian Railway in a series of battles that earned the CL the grudging respect of their enemies.

President Wilson, under pressure from the Allies to join in anti-Bolshevik military operations in Russia, finally decided to dispatch the American Expeditionary Force, Siberia, to Vladivostok. Composed of nearly 8,000 men of the 27th and 31st Infantry along with volunteers from many other regiments, the AEF Siberia arrived in Vladivostok in mid-August of 1918, with their commander, General William Graves, arriving in early September.

Contrary to the expectations of the Allies, General Graves believed that his mandate only extended to securing United States military materiel, and aiding the evacuation of the Czechoslovak Legion. Throughout his command, he was pressured by Allied commanders in the region from the British, French, and Japanese armed forces to take the fight to the Bolsheviks, but he remained steadfast.

His troops saw combat, nonetheless. The AEF Siberia did eventually go into battle in a defensive posture, repelling Red Army, partisan, and Cossack raids on the Trans-Siberian Railroad and the supplies at the Vladivostok docks. They also experienced strained relations with their Japanese allies, intent on seizing the resource-rich lands of the region for themselves.

It was the land itself that proved to be the greatest obstacle for the AEF Siberia. Prior to their deployment to Russia, most of the AEF Siberia were stationed at the United States bases in Manila, the Philippines. The transition from a tropical climate to the arctic conditions of Vladivostok proved too much for both soldiers and equipment. Limb amputations due to frostbite were common, the regiment's water-cooled machine guns froze and became inoperable, and their horses, bred for temperate conditions, proved to be less than useless, and died in vast numbers.

The goals of the expedition also proved fruitless, doing little to aid in the evacuation of the Czechoslovak Legion, with the war materiel mostly winding up—after the AEF was evacuated from Vladivostok—in the hands of the Red Army anyway.

Yet they endured, fighting on until they were at long last brought home in April of 1920, almost a year and a half after the signing of the Armistice. The force lost over 200 men, mostly to frostbite and exposure.

At the exact same time on Russia's Western coast, the 339th Infantry was arriving at Archangel'sk with much the same mission: secure Allied war materiel. Composed largely of soldiers from Michigan, the 339th had originally been destined for France but was diverted at the last minute as part of the North Russia Intervention, the Allies' active attempt to subvert the Bolshevik government and bring the White Russians to power.

Placed under command of the British, these soldiers were given a much more active mandate to fight the Bolsheviks. They were used in offensive operations almost from the beginning of their deployment until the front of battle along the Dvina River and Vologda Railroad became hundreds of miles long and a supply nightmare. With the onset of winter, the 339th adopted a defensive posture. They would remain so, suffering as their brothers in arms to the East did from frostbite, partisan attacks, and the Spanish flu, for another six months.

With the signing of the Armistice, the 339th expected that they would soon be sent home. But the White Sea port of Archangel'sk had frozen over for the winter, preventing demobilization, and between that and the grueling nature of their duties, morale began to steadily decline. Letters home complained of privation, lack of food, and the cold, demanding to know what they were still doing in Russia if the war was over.

In the end, it was the tide of public opinion that ended U.S. intervention in Russia. Family members of the 339th began writing furious letters, to their congressmen, to President Wilson, and to every newspaper that would publish them, demanding to know why their boys were still fighting when the war was over. Nevertheless, the Wilson government was slow to act, and mutiny began to break out among the 339th. Court-martials were threatened (but never manifested), and in February of 1919 Wilson finally signed orders to retrieve the 339th, who came home after a dramatic rescue involving an icebreaker in April 1919, a full year earlier than their unfortunate fellows in the East.

During that journey home, the 339th decided to adopt the moniker "Polar Bears," and were granted permission to add a polar bear patch to their uniforms. They retain that honor to this day, and a memorial to the expedition stands in Troy, Michigan. Largely forgotten by history, the Polar Bears and the AEF Siberia provide an important clue to U.S.-Soviet relations: the narrative usually focuses on the U.S. and Soviet alliance during the Second World War, and adopts a bewildered tone at the Cold War split which followed thereafter. But a thorough analysis of the Great War shows us that Soviet hostility toward the United States was well-founded from long before and is no mystery.

THE RED SUMMER

The events of the spring, summer, and early fall of 1919, which are collectively known as the "Red Summer," occupy that strange hinterland of our national consciousness where we banish things we would rather not think about. Shameful events in our past are often not well-known, and the Red Summer is manifestly both shameful and virtually unknown.

Following the Bolshevik Revolution, the Wilson government became increasingly alarmed about the possibility of a similar uprising occurring in the United States. The revolutions in Germany, Hungary and elsewhere did nothing to assuage their concerns. Combined with vastly increased strike activity during and after the war, a series of bombing campaigns against prominent social and political leaders committed by anarchists, and near-hysteria in the press, the patriotic political climate surrounding the Armistice soon devolved. Jingoism, racial hatred, and violence became the order of the day.

The nation's anti-German sentiment transmuted seamlessly into anti-Bolshevik sentiment; the return of hundreds of thousands of soldiers to civilian life placed a great strain on an economy ill-prepared to receive them. This volatile combination led to a scenario almost unimaginable now: the lynching of United States servicemen in uniform.

African American soldiers, already forced to serve under humiliating segregated conditions that relegated most of them to non-combatant duties, returned from the war to find their service actively denigrated. African American men had responded to W.E.B. Dubois' call to serve in the segregated military in vast numbers; 380,000 enlisted, and despite their treatment by the United States military, their experience among French civilians was very different.

Treated with unaccustomed respect and dignity in France, many African American servicemen realized that the racially divided society of the United States was not the only way things could be done. They returned home with a greater sense of purpose and determination to improve conditions for their community.

That purpose met with incredible resistance, especially in the South. On thirteen separate occasions in 1919, African American soldiers were lynched, in each case because they had the audacity to continue wearing their uniforms. Why this was regarded as such an affront to white people is almost incomprehensible to the modern mind; the words of an editorial entitled "Nip It In the Bud" from a Louisiana newspaper of 1918 are a window into that kind of thinking, and they are chilling: "This is the right time to show them what will and will not be permitted, and thus save them much trouble in the future."[2]

What would be permitted was violence. Beyond lynching, anti-African American violence was sparked in multiple cities owing to the use of African Americans as strikebreakers. Particularly vulnerable to unemployment and ironically attempting to flee violence, many black families left the South for the industrialized North and Midwest. Often unable even to join the unions, their only recourse was to act as scabs to gain any employment at all. This provoked rioting in DC, Chicago, and many other cities.

This was not the first time anti-black rioting had occurred in the United States. It was, however, the first time that African American citizens fought back. In nearly every instance, African Americans, supplemented by their returned veterans, refused to be dominated by racial violence and stood up for themselves, which often provoked reprisals more drastic than the initial attacks. In one particularly shameful incident in Washington DC, white servicemen joined in with the white supremacists in hunting down and killing African Americans over a four-day period, owing to a rumor of a black man's arrest for the rape of a white woman.

And yet, despite these terrible affronts, African Americans would once again serve with pride and distinction in the next global conflict, enlisting in vast numbers for the Second World War. Over 1.2 million served their country in a still-segregated military.

In conclusion, the end of the Great War for the United States cannot be given a date or a time. It continued for many months after the Armistice and its repercussions were felt throughout American society very unlike the absence of them today stemming from our current wars. The lessons of the Great War continue to inform our understanding of an America that has always strived to be a good place, always cast itself as a paradise and an ambassador of peace. When we do not reach these ideals, we must ask why. Historians must never shy away from the hard questions of our past, nor embrace the tidy and convenient answers that never tell the whole story.

Rob Bokkon lives with his wife, son, and two cats in a college town in Kentucky.

Notes

[1] Patti Minter, "To Rob Bokkon," n.d.

[2] E. Robinson and M.E. Robinson, "Nip It in the Bud" in *True Democrat,* Dec. 21, 1918, *News about Chronicling America* RSS (Bayou Sara Pub. Co.), accessed January 20, 2022, https://chroniclingamerica.loc.gov/lccn/sn88064339/1918-12-21/ed-1/seq-2/.

Brest-Litovsk:

Eastern Europe's Forgotten Father

ADRIAN BONENBERGER, veteran

One of the greatest ballet dancers of the 20th century—perhaps ever—Vaslav Nijinsky was born to Polish parents in what is now Ukraine, was at the time Imperial Russia, and for a brief six-month period, was a puppet-state of Imperial Germany. He is supposed to have thought of himself as Polish. He married a Hungarian woman and raised his children as Polish. He spoke Russian.

I find Brest-Litovsk extraordinary because of the assumptions that Germans made about the ways populations and peoples worked. It causes us to consider the meaning of identity at the dawn of the 20th century. Although Germany had expended great military, financial, and political capital in annexing or expanding its influence at Russia's expense—a fact that they hoped would be to Germany's financial and political benefit—to the people who lived within that territory, Germany's sovereignty must have seemed like a distant abstraction, a matter of architecture and linguistic preference. As intellectually similar and similarly modern as even the most cosmopolitan people of that time, they were nevertheless different—there was a kind of existential freedom to choose one's path, to pick an identity, rather than having it forced upon one.

Traveling in the ruins of Ukraine's east today, the shuttered factories and empty industrial streets, it's impossible to overstate how much ethnic nationalism has played a determinative role in shaping how people think about themselves, and how they think about others. Countries are now in places where they haven't been before for many years—Ukraine, Belarus, Poland, Finland, the Baltic States—Germany's sundering of European Imperial Russia not only helped lay the groundwork for those countries, but also for how people think about themselves within those countries.

In this respect, the Treaty of Brest-Litovsk is the moment the old world falls apart and creates space for the new to arise. Imperial Russia had resisted giving Ukraine its autonomy, but after Brest-Litovsk and a brutal war, Russia ended up making Ukraine and Belarus independent S.S.R.s in their own rights. It did the same before and then after World War II with the Baltic States.

There is a kind of story about individuals that gets written during the late 19th century and early 20th century that is recognizably modern in the sense that people have interior dialogues and make choices about their lives independent of their external circumstances; those characters, like Hardy's Tess or Jude or Hemingway's male protagonists, are crushed by the systems that exist around them. The stories I love that come later in the 20th century—those by David Foster Wallace, Margaret Atwood, Raymond Carver, and Denis Johnson—all include characters defined by their circumstances in ways that earlier characters would have been trapped. In a world where one's birthplace renders something vital to identity, choice takes on a very different quality than it did in the Europe that predated WWI.

There's an apocryphal story about Nijinsky as an old man. Having apparently lost the power to speak Russian during WWII, he encountered Russian soldiers in post-war Vienna dancing and playing folk instruments. He joined them, and astounded them with his dancing prowess, then settled down to drink with them and carouse, regaining his ability to speak. I like this story very much, because of what it says about what's important to people, or, at least, what's important to the type of person Nijinsky was—a cosmopolitan person with cosmopolitan values.

WHERE THAT WORLD OF POSSIBILITY ENDED

The future of Eastern Europe was decided on March 3, 1918, in the town of Brest-Litovsk. Imperial German and Bolshevik Russian negotiators had met repeatedly in the preceding weeks, each seeking to improve their position. With German forces advancing and Russia's Army on the verge of disintegration, Russia did the unthinkable, signing away a quarter of its population and a great deal of its industry. The Baltic States, Ukraine, and Belarus were all separated from Russia; liberated or annexed into the German Empire. It was seen as the best possible outcome for Germany, the worst for Russia, and, at the time, seemed like it might give Germany the power it needed to push through and end the War victorious.

Things didn't work out that way. Germany was defeated less than six months later, and the Treaty almost immediately abandoned or revoked by all participants. With such a short practical lifespan, it might seem like the Treaty of Brest-Litovsk is just one of those oddball moments in history, a wistful could-have-been for Germany, and a nightmare averted for everyone else. This overlooks the profound impact the treaty had on defining the basic outline of modern Eastern Europe, presenting new opportunities for populations eager to exist outside Russian borders.

NEGOTIATING FROM A POSITION OF WEAKNESS

Germany's plan with the treaty was simple: it held all the trump cards—a unified country, a powerful military, and a clear set of priorities—and it intended on forcing a swift agreement before the Bolsheviks could collect themselves. Lenin's fragile new coalition, on the other hand, was fighting a civil war, and, besides, a substantial portion of its appeal to the Russian Empire had been a swift halt to hostilities, as war was a tool of capitalist exploitation. Russia had very little room to breathe.

It should go without saying, but the thoughts and opinions of people native to and living in the countries in question were not consulted—or, at least, not in good faith. Ukraine and Belarus, both liberated as countries independent from the Austro-Hungarian Empire or Germany, were not aware that this liberation would mean economic or military dependence on their Western neighbors, but that is precisely what Germany intended in its terms of peace.

What seemed like strength, however, ended up being the opposite, and vice versa. Germany insisted on and received its demands, and then found itself forced to garrison its new territories with nearly a million soldiers, overstretching its military and potentially robbing itself of how to force a conclusion to the war before Americans could arrive.

The garrisoning of soldiers did not come without its own risks—although the Bolsheviks had reluctantly agreed to terms that were widely viewed as humiliating, and were no longer in open conflict with Germany, that did not stop local Bolsheviks or their socialist allies from attempting to radicalize Germany's soldiers. Furthermore, ethnic nationalists in Ukraine and Belarus felt that they had been misled and began small-scale insurgencies. What had seemed originally like a boon to the German war effort quickly turned into a cultural and logistical nightmare.

LONG-LASTING EFFECTS

While these immediate effects of Brest-Litovsk were playing out—soon to be overcome by Germany's loss of the war—the longer-term implications of the treaty are most relevant today. Chiefly, the establishment of a clear legal precedent for an Eastern Europe to exist independent of Russia, in a basic form that endures to this day, along the model of countries to its West, but also the basis for that precedent being rooted in the ethnic, historical, and linguistic identity of people inhabiting those places. Although "Ukrainian" and "Belarusian" culture were recognized as such long before Brest-Litovsk, they were not viewed as self-sustaining cultures. On the contrary, those identities made Ukrainians, Belarusians, Estonians, Lithuanians, Latvians and Poles exceptionally dangerous to

Germany and to Russia. Germany facilitated the breaking of that long standing tradition of subjugating Eastern European peoples, subsuming them into larger nations.

The most immediate consequence, sadly for the Central Powers after their defeat, is that the same principles were applied to Germany and Austria-Hungary, and both countries were partitioned to create smaller nations (Germany much less so than Austria-Hungary, which was destroyed). What had seemed like a good idea when applied to others was catastrophic when applied to them. And the scope of the Treaty of Versailles would likely not have been as sweeping or severe had not the Treaty of Brest-Litovsk been hovering over the proceedings.

Looking at a map of Eastern Europe today, it's like the map drawn up by negotiators sitting around tables about 100 years ago, dreaming of a German empire, or of a worker-led socialist revolution.

A final point that's worth making about Brest-Litovsk, Eastern Europe, and how it came about: the shape of those countries liberated or annexed by Germany were not put into their final form until 1946, for the most part, with pieces of Ukraine, Belarus, and Poland shifting and changing repeatedly. In the case of Ukraine, if Russian occupation holds, one may be able to say that the shape of its country changed last in 2014.

But most of those changes were imposed externally, created by treaties or agreements by those larger powers. Few countries in Eastern Europe achieved independence or identity through war, or on their own terms, for better or worse. Tim Snyder makes a compelling case in his book *Reconstruction of Nations* that the closest "positive" national movement occurred in Poland's handling of the dissolution of the Warsaw Pact and the end of the U.S.S.R.[1] It took the initiative, unprompted, to relinquish outstanding potential claims on Ukrainian territory. Their action was a kind of positive national determination that helped set the stage for further positive relationships between Eastern European countries. This was during a time when other areas, notably the former Yugoslavia, fell into extreme violence. There was no reason to expect that would not be the case in Eastern Europe, especially given the two wars fought between Poland and Ukraine between 1918-19 and then 1943-46.

The Western European attitude toward Eastern Europe as a place on which to impose one's will and one's destiny—a place to colonize, rather than a place of intentionality and agency—is longstanding and may even go back to Roman times. The quasi-racist perception of Eastern Europeans by British, Italians, and Germans has driven past imperialist projects and also exists today, driving populist, nationalist, xenophobic sentiment.

Why does this matter? If one imagines a Prussian Junker signing a paper on a beautiful oaken-wood table, assuming the notion that an action like the establishment of a subordinate and subordinated Ukraine are in the past, well, there's an argument to be made that this moment is much closer to the present time than many might realize.

EASTERN AND CENTRAL EUROPE IN POPULAR AND PERSONAL IMAGINATION

Eastern and Central Europe, the two places affected by Brest-Litovsk and Versailles, respectively, both experienced life within an empire, followed by the destruction of that authoritative social order. It shouldn't be surprising that the type of stories and storytellers that emerge from Eastern and Central Europe from that late 19th to early 20th century period are similar, depending on surrealism that sometimes wanders into fantasy, as in the works of Kundera, Bulgakov, Kafka, and even Vasily *Grossman's Life and Fate.*

Those stories shape our memory at that time, as per Wes Anderson's *The Grand Budapest Hotel,* which accurately reflects how some remember the lives and lifestyles of people in what could easily be Budapest, Brasov, Odessa, Warsaw, or Vienna. The disorder of the 1920s and 1990s encourage populists and pastoralists alike. As I walked through the ruins of Soviet-era sanatoriums in Eastern Ukraine, or through Romanian villages where decayed 13th-century fortified German churches sit side-by-side with cheap, laundry-bedecked communist-era housing, I was surrounded by the ghosts of those periods of our shared past—a collective past, an imperial past, and the past that gave rise to the relative independence and individuality of a post-modern world, as well as the literature that arises naturally from it.

Adrian Bonenberger is a writer who lives in Connecticut. His war memoirs, *Afghan Post*, were published in 2014. A collection of short fiction, *The Disappointed Soldier and Other Stories from War* was published in July of 2021. He's a co-editor with *Wrath-Bearing Tree*, an online literary publication.

Notes

[1] Timothy Snyder. "Patriotic Oppositions and State Interests (1945-89)", in *The Reconstruction of Nations* (New Haven: Yale University Press, 2003).

The Balfour Declaration:
An Alternative History

SIMONE ZELITCH

We're living in an alternative-history moment. Given the current political and cultural climate, it's no coincidence that many of us are in full flight from the present. Instead, we look over our shoulders, endlessly revising the past.

After President Trump's recognition of Jerusalem as the capital of Israel in 2017, it's time for one more revision: What if there had been no Balfour Declaration?

Balfour addressed his declaration to Baron Lionel Walter Rothschild whose family name often appears in antisemitic conspiracy theories. November 2017 marked the centennial of that document, which begins: "His Majesty's Government views with favour the establishment in Palestine of a national home for the Jewish people and will use their best endeavours to facilitate the achievement of this object."[1]

The Balfour Declaration was far from altruistic. As England, France, and Italy prepared to divide a defeated Ottoman empire, England wanted access to the Suez canal, a link to their colony in Egypt, and although the endorsement certainly grew out of respect for the Zionist project, I admit to finding it difficult to ignore the influence of a familiar anti-Semitic trope: Jews have money, influence and power, and it makes good sense to get them on your side.

I could write far more about division in the world Jewish community about a homeland in the territory of Palestine, or indeed division among Zionists themselves. Indeed, far more could be written about the many promises made to many of the people in the region who hardly expected their homelands to be divided like figgy pudding. These subjects have been discussed so thoroughly that it's hard to get through the dense competing ideologies.

What haunts me instead is this: Jews, Muslims and Christians lived in Jerusalem, Jaffa, and Hebron for generations before those cities were part of a country with fixed borders; those cities and the surrounding countryside were under the jurisdiction of the Ottoman Empire, with a government so inefficient that it might as well have not been there at all. What was life like for Jews in the area that Balfour referred to as Palestine?

This is one of the questions addressed in Menachem Klein's 2014 book, *Lives in Common,* which focuses on Arabs and Jews in those three cities before the British Mandate, and goes on to explore how relationships were transformed by emerging national identities in the years that followed. The early chapters of the book describe a time when Arab and Jew weren't separate, contradictory categories; in fact, there were Arab Jews, indigenous and integrated into their communities, without distinct national aspirations.

Klein gives us tantalizing portraits of Jewish musicians playing at Muslim weddings, of Jewish and Muslim families mingling in common courtyards, of bathhouses that date from the Mamluk period used as by Jewish women as mikvahs, of the 451 Arabic words that found their way into Yiddish, of cafes in cosmopolitan Jaffa where Arabic mixed with Ladino, and men would gather to play dominoes, talk politics, and read the local newspaper, *Falastin.*

Of course, these "lives in common" are complicated by real prejudices that pre-date the Balfour Declaration.[2] Under the Ottomans, Jews were second-class citizens, and Muslim and Christian Arabs made clear distinctions between "local" (Arab or Sephardic) Jews and "Muscovites," who were considered invaders. They were Ashkenazi Jews under the protection of European embassies, whether they were strictly religious and dependent on charity, or secular Zionists farming on property they bought from absentee landlords in Damascus.

Yet it's worth considering: if Britain hadn't supported Jewish nationalism, would Arab nationalism have turned in the direction of antisemitism? What if the territory we call Israel or Palestine was part of an Arab country with borders that may not in any way resemble the borders we see on maps today? Would Zionists continue farming and reviving the Hebrew language there? If Ottoman restrictions remained, would those Zionists unite with Sephardim, Ashkenazim and Arab Jews to demand full citizenship in this imaginary country?

Of course, this speculation may, in the end, be just as likely or unlikely as the vision asserted in the Balfour Declaration, particularly when we read beyond the initial phrase and find that the document endorses its homeland for the Jews with caveats: " . . . it being clearly understood that nothing shall be done which may prejudice the civil and religious rights of existing non-Jewish communities in Palestine . . . "[3]

Is it possible for any "national home" to meet the Balfour Declaration's standards? Israel can't now, and year by year, that goal feels more remote. Yet Israel is not alone. Today we struggle with the consequences of national homes. The collapse of the Soviet Union, much like the collapse of the Ottoman Empire, left a map to be redrawn.

So, what about Jerusalem, with its gold stone, cramped alleys, fanatics, and bedazzled tourists? That city is, of course, a magnet and an incubator for these prejudices. It's tempting to wish Jerusalem was blasted into orbit as a satellite, a kind of golden moon, an aspiration.

We can wish the same of nations—that identity, language and culture could develop without armies and borders. Some influential Zionists like Martin Buber and Ahad Ha'am explored this possibility. We can't revise history, but we can revisit missed opportunities and see if they lead us away from the dead end we've clearly reached today.

As Zionism's founding spirit, Theodor Herzl, once wrote, "If you will it, it is no dream."[4]

Simone Zelitch is the author of five novels, most recently *Judenstaat* (PM Press 2020), an alternative history about a Jewish State established in Germany after World War Two. Her work has been taught in colleges across the country, including at the University of Miami in a class called "Bad Jews." She teaches at Community College of Philadelphia and hopes to travel to Israel and Palestine soon to research a new novel about the political implications of archaeology. Find out more about the author and her work at www.simonezelitch.com

Notes

1 "Balfour Declaration: Text of the Declaration," Text of the Balfour Declaration, accessed January 21, 2022, https://www.jewishvirtuallibrary.org/text-of-the-balfour-declaration.

2 Menachem Klein. *Lives in Common: Arabs and Jews in Jerusalem, Jaffa and Hebron* (London: Hurst & Company, 2014).

3 "Balfour Declaration."

4 Theodor Herzl. *Old New Land = Altneuland* (New York: M. Wiener Pub., 1987).

A New Look at Anne Frank, Her Father, and WWI Through Literature and History

DAVID R. GILLHAM and PETER DE BOURGRAAF, PhD

***

The Literary Perspective: Otto Frank And The Sergeant
DAVID R. GILLHAM

During the First World War, Anne Frank's father, Otto Frank, served in the Kaiser's army as did over 100,000 German Jews. His maternal family had roots in his native Frankfurt that stretched by centuries. Otto himself studied at the University of Heidelberg, and eventually, he went to work in the family-owned bank. He was assimilated and thought of himself as German as much as Jewish.

During the war, he enlisted and was assigned to the Imperial Field Artillery as a *Kanonier*. In 1916 he survived the carnage of the Somme unscathed, and likely witnessed the introduction of armored warfare onto the battlefield when Mark I tanks of the British Machine Gun Corps Heavy Branch lumbered into history at the Battle of Flers–Courcelette. During that same year he was promoted to the rank of *Feldwebel*, sergeant-major, and in the next year was awarded a field commission as a junior lieutenant, a *Leutnant,* for his courage during a dangerous reconnaissance of enemy artillery. He ended the war as an officer, Lieutenant of Reserves Otto Heinrich Frank, recipient of the Iron Cross, 2nd Class.

Twenty-seven years later, he would be hunted as a criminal Jew by the German occupiers of the Netherlands, hiding out with his family and friends in the rear annex of his office building on Amsterdam's Prinsengracht. And, his rank as a young officer in the Great War would have a small but ironic impact on the final day of their time in hiding—the day of their betrayal and arrest.

On August 4, 1944, Karl Josef Silberbauer of the SS security service, the Sicherheitsdienst or "SD"—also known as the "Grüne Polizei" because of the color of their uniforms—was dispatched to a building in Amsterdam's Canal Ring where he arrested the eight Jews found in hiding there.

How did this come about?

It's been recently proposed that the purpose of the raid was actually to investigate suspected black-market activities and that the SD men had simply stumbled onto the Frank family and their friends. But the more generally accepted version of events is as follows: Silberbauer received orders from his superior, an SD officer named Julius Dettmann, based on a telephone tip that there were Jews at Prinsengracht 263.

Who called in this tip? If indeed a call was placed? There are many theories but none of them proven. An SD chief, Willy Langes, maintained after the war that procedure would have dictated the steps taken after any such tip. That any such telephone tip would have by necessity come from a known and reliable informer in order for action to have been ordered on that same day. Julius Dettmann committed suicide after the liberation, so if he did receive a call, and if he did, in fact, recognize the caller as a reliable informant, he then took this information with him to his grave. Karl Silberbauer had not bothered to ask about details. It was sufficient that he had been given an order, so he followed it without further question.

Silberbauer himself was Viennese by birth. He had been a soldier too in the last war, and then a police officer. In 1938 Austria was formally annexed into the German Reich during the Anschluss, after which Silberbauer joined the Gestapo in Vienna. In 1943 he transferred to the occupied Netherlands as a member of the SD—the SS Security Service or Sicherheitsdienst. There he was assigned to

Sektion IV B4—the so-called "Jewish Desk." His rank in the SS was equivalent to that of a sergeant. When he arrived at the Prinsengracht 263 that morning in August, he was the only member of his squad in uniform, accompanied by a cohort of Dutch plainclothes men, collaborators employed by the security police for just this sort of work.

He was an unremarkable man. Years later Ernst Schnabel interviewed Miep Gies, Otto's office secretary and one of the primary Dutch helpers providing care for the Franks and their friends while in hiding. She characterized Silberbauer as a man who "looked as though he might come around tomorrow and read your gas meter or punch your streetcar ticket."[1] And certainly, the raid on Prinsengracht 263 was routine for Silberbauer. He had arrested many Jews in hiding by now. The inhabitants of what was to come to be known as Het Achterhuis—"the House Behind"—were given five minutes to pack up their belongings. Standard procedure.

But then something happened. In a post-war interview, Silberbauer claims that Otto Frank introduced himself as a veteran of the First War. According to Otto, however, it was his old footlocker that caught the SD man's attention. The gray ironbound chest came from Otto's time in the service bore his name and rank. Lieutenant of Reserves Otto Frank. According to Otto's post-war conversation with the author Ernst Schnabel, when Silberbauer spotted the object, his entire demeanor altered.

He "stared down at the chest that stood between bed and window, and exclaimed 'Where did you get this chest?' 'It was my own,'" Otto had replied, and then added, "'I was an officer in the First War.'"

Silberbauer was astonished. "The man became increasingly confused," Otto told Schnabel.[2]

Somewhere inside him, this sergeant must have snapped to attention in the presence of a superior officer. He didn't understand why Otto had not simply registered himself with the Nazi-run Jewish office as a veteran and former officer. The SD man insisted vehemently that Otto and his family would have treated "decently." That they would have been sent to Theresienstadt in Bohemia—one of the so-called "paradise camps," where the Nazis had even produced a ludicrously staged film of busy, well-fed camp inmates populating the Theresienstadt camp for use as propaganda in an attempt camouflage the truth about Sobibor and Auschwitz-Birkenau.

Otto's response to Silberbauer was silence. "'I said nothing,'" he told Schnabel. "'Apparently [Silberbauer] thought Theresienstadt a rest camp, so I said nothing. I merely looked at him. But he suddenly evaded my eyes, and all at once the perception came to me: Now he is standing at attention; if he dared, he might very well raise his hand to his cap in salute."[3] The discovery of Otto's former rank left the SD sergeant flummoxed. So, he allowed them extra time to pack. "Take your time!" he called out several times. "He shouted these same words to us and to his agents," Otto recalled.[4] This was Silberbauer's single concession. Meanwhile, the man had already made himself busy stealing what meager valuables were to be had. To carry his booty Silberbauer had confiscated Otto's briefcase, simply dumping its contents onto the floor. The contents included Anne Frank's diary, her notebooks, and her collection of loose-leaf pages, all left behind in the wake of the arrest. It was Miep Gies who collected the books and pages and kept them safe in a desk drawer waiting for the day of Anne's return.

After the war, Karl Silberbauer served fourteen months in prison, not for arresting Jews and sending them to their deaths, but for having assaulted some Austrian Communist Party members. Afterward, he returned to the Viennese police force and was recruited by West German intelligence as an informant. It wasn't until 1963 that the camp survivor and Nazi hunter, Simon Wiesenthal, tracked him down and exposed *Inspektor* Silberbauer as the man who had arrested Anne Frank and her family. He was suspended without pay pending a judicial investigation and a disciplinary hearing.

But ironically, Otto Frank played a part in determining Silberbauer's fate. Anne Frank's diary had already become an international phenomenon by this point. According to Wiesenthal's biographer, Joseph Wechsberg, Otto's maintained that Silberbauer may have been condescending, but not actually abusive to his captives. Instead, Otto placed blame on the unknown person who had betrayed them to the SD. Owing much to Otto's statements on the matter, Silberbauer was exonerated. The judicial investigation dissolved, and the internal disciplinary hearing concluded with his suspension lifted. He returned to the police force, assigned to a desk job indexing criminal records. After his retirement, he was provided with a government pension. Benefits which, according to the Simon Wiesenthal Center, were accruing during his time with the SS security service in occupied Amsterdam. He died in 1972, age sixty-one.

But before his death, he accepted an interview from a journalist, Jules Huff, who was reporting for a Dutch newspaper in The Hague. According to Carol Ann Lee's biography of Anne Frank, during the interview, the former SD sergeant was quoted as saying: "I bought the little book last week, to see if I am in it. But I am not."[5] He also insisted that not only was the entire investigation of his past designed to "butter up the Jews,"[6] but that Otto had himself brought fate down upon his family by going into hiding in the first place. Maintaining till the end that it was the victim's fault, not the fault of the perpetrators. And of course, he pleaded ignorance. "We never knew what was happening to the Jews. I've no idea how Anne Frank came to write in her diary that Jews were being gassed."[7] The BBC was advertising this fact as early as June of 1942. Anne wrote of it in her diary in October of that same year.

Of Otto, he said, "If Frank hadn't gone into hiding, nothing would have happened to him. For me, the whole thing was dealt with in an hour. The eight were put on a transport and I never bothered about them again."[8]

Miep Gies, the helper who rescued the diary from the floor, died in 2010 in North Holland at the age of one-hundred.

Otto Heinrich Frank died in Switzerland in 1980 at the age of ninety-one.

Otto's two daughters, Anne and Margot, had died of typhus thirty-five years earlier, between February and March of 1945 in a concentration camp called Bergen-Belsen.

Anne Frank's diary, however, lives on.

David R. Gillham studied screenwriting at the University of Southern California in Los Angeles before transitioning into fiction. After moving to New York City, Gillham spent more than a decade in the book business, and he now lives with his family in Western Massachusetts. He is the author of *Annelies: A Novel of Anne Frank.*

Notes

[1] Ernst Schnabel, *Anne Frank: A Portrait in Courage*, trans. Richard and Clara Winston (New York: Harcourt, Brace and Company, 1958), 133.

² Schnabel, *Anne Frank*, 135-136.

³ Schnabel, 136.

⁴ Schnabel, 136.

⁵ Carol Ann Lee, *Roses from the Earth: The Biography of Anne Frank*, (London: Viking, 1999), xv.

⁶ Lee, *Roses from the Earth*, 245.

⁷ Lee, 246.

⁸ Lee, 246.

The Historical Perspective:
World Wars Mirrored in the Secret Annex

PETER DE BOURGRAAF, PhD

Amsterdam and Berlin-based, I live what you might call a trans-European life. I am a Dutchman who works and researches mostly in the German language as a freelance historian but also self publishes in English. My work with Amsterdam's Anne Frank House inspired me to pursue Central European history further. I'd already been working on this subject for a quarter of a century, especially addressing the neglected parts about the ways WWI's contended closure resulted in another world war and the Holocaust. The powerful words of Anne Frank's famous diary motivated me to study Otto Frank, her father, and his involvement with WWI. The conclusions from my research include a call for an international scholarly task force to deal with what I ascertain to be an untold, century-old story.

The outstanding remembrance work of the United States World War One Centennial Commission (WWICC) clearly demonstrates that this war did not end in 1918. The Treaty and "League of Versailles" formally terminated a nearly five-year-old belligerency. While European Great-War mnemonic performances saw a full stop at the end of November 2018, it was not until the end of next year before the WWICC's mission was completed. It even went on after the Versailles anniversary of June 28, 2019. While carefully observing the ceasefire centennial including the Paris Conference daily, there was zero reporting about the all-American party at this *lieu de memoire*. The headlines of the Brexit process that, most notably, took as long as the warfare until 1918 seemed to preoccupy the historically ignorant and disaffected Europeans.

In fact, the 1914 World War has, in my opinion, no clear end at all, except for the German-American Peace Treaty of August 1921 when the Armistice initiators found themselves back together. That ended three decades later in 1945 Berlin/Nagasaki and amounts to what I would call a second thirty-year war. The term was coined by Winston Churchill, among others, referring to the 1618-1648 Thirty Year's War during which Germany's towns and landscapes were devastated and its territorially-scattered populations decimated.

I have tried to visualize the story of the 1919 early winter days in Paris and to evaluate its recent anniversaries. Ever since my inspiring experience with the Anne Frank Foundation, I had always wondered about when the lives of the then just-demobilized Western Front comrades-in-arms like Otto Frank and Adolf Hitler grew apart. Austrian and German soldiers like them stood shoulder-to-shoulder in horrendous warfare that had lasted for almost three years when the United States decided to join the enemy's side. Germany's chances of winning the war were virtually reduced to zero by this April 1917 groundbreaking development.

These two soldiers' mortally divergent post-Western Front paths go back to the largely forgotten events of the opening of the Versailles Conference. The British appeared twice as strong as the other participants. Not at all bothered by the prevailing and just prolonged November 1918 Armistice agreement, they did away with those who protested a British Imperial Delegation. Independence seeking colonists from South-Africa ended up leading this second delegation under the Union Jack and their fellows from Australia added to the extremely nationalist outlook. Severely damaged Belgium was left far behind, and David Lloyd George with the Associated United States of America, etc., came out in front as the leading delegations. The Empire and British delegations went on to dictate imperialistically motivated and orchestrated demands to both friend and foe. Following the ceasefire agreement, the German underdog would naturally have joined the party at some time, but the Allies and U.S. President Woodrow Wilson decided to block this and go unilateral until the end.

This is not usually what schoolbooks and history lessons across the world have taught us. These keep telling us how the world glorified Wilson and his brainchild, the League of Nations. Furthermore, the impositions of French President Georges Clemenceau deprived another newborn the hungering Weimar republic–from livelihood.

In defiance of this Anglo-Saxon narrative, which dominated historical consciousness around the world even after Germany's surrender in 1945, millions of Central Europeans such as Frank and Hitler experienced it differently. In general, the conditions and consequences of the mutually acceptable Armistice fell beyond the scope of their everyday lives. For instance, Alsace-Lorraine had been taken back by France. Following the British twin delegations' coup in Paris, the French were not reluctant to claim more German territory, i.e., Rhineland provinces. These claims were blocked by the British Empire's leadership of the Entente because, as my research shows, they were focused on annexing the German colonies. As far as the triumphant motherland was concerned, the whole of Germany's overseas empire was to be absorbed. In what is also known as a "reverse Alsace-Lorraine," the French Rhineland claims were thereby virtually replaced with something that may be regarded as the birth of an even more frightening Alsace-Lorraine overseas even though annexations in Asia, Africa or the Pacific were ruled out by the effective ceasefire provisions.[1]

Thus, historically, Wilson's leading position in establishing the League of Nations was undermined as a consequence of the British—a "British" colonial and imperialist coup. Now, a century after the events of Paris and Versailles, is it not a widely accepted view that the U.S. Senate saw this act as a crucial argument when it canceled the ratification of the Treaty of Versailles on November 19, a year after the end of hostilities. As a matter of fact, the sustained acceptance of the Monroe Doctrine argument is congruent with the recollection of these forgotten and ignored events. I would argue that the British and their nationalist colonists got from the indispensable President and remaining armistice guarantor what they secretly had been aiming at: the Empire's worldwide expansion, unchecked rule and unchallenged leadership of a novel international organization. As of November 1919, the hollowed-out League was firmly in British hands. The post of Secretary General was occupied by a Brit who would hold it almost as long as the Weimar Republic existed.

And, most illustrative of this 21st-century tale of the Armistice's crumbling and the failure of the Paris peace is what happened about twenty-five years later in a ten-minute window when the "Secret Annex," the Frank family's hiding place, in Amsterdam was raided. After more than two years, the *Sicherheitspolizei* had discovered the hiding place of the now-famous family of Anne Frank and their four companions. On August 4, 1944, the deportation operation was led by Mr. Silberbauer, a sergeant and native from Austria. In contrast to the obvious disinterest of his Dutch assistants, a seemingly unimportant object caught the thirty-three-year-old's attention. Mr. Frank did not need to confirm that the spotted WWI army crate was his, because his name and rank of lieutenant were on it. Suddenly, the agonizing moment made way for relaxation. The German world-war Jewish veteran in hiding was instantly granted a ten-minute time frame in order to ensure a neat departure towards deportation for both his family and the others.

This is why I think that an interesting follow up to a visit to Amsterdam's Anne Frank House would entail reading my post-WWI-anniversaries investigative and evaluative article, *Decolonizing the "League of Versailles." The Sleeping Beauty of Commemorative Culture* (2022). It shows that the above-mentioned imperialist Paris/Versailles diktat stood in the way of a normal and peaceful life in the heart of Europe. Generations of commensurately humiliated and ostracized Germans would be followed by the deadly victimization of the Jewish people and others persecuted during the world war's continuation in the 1940s.

The indescribable horror of world war would wait a little longer. In no way am I arguing that any radically revisionist Germany was justified in implementing genocide. Any mitigation of these horrors will be out of the question. However, a revision of post-WWI is needed and justified to restore a sound community of peoples and nations. Despite the existence of Weimar Republic and other established democracies, the above-mentioned derailments persisted.

The devastating diktat paved the way for a terrible dictatorship that echoes George Orwell's paradigm: "an end with horror is better than a horror without end."[2] Both of these ends fell upon the only nation truly incriminated, stigmatized and punished for the Great War.

Not before this intricate story will be translated and told over and over, this nearly seventy-year-old mainstream history, which says that the Paris conference was about peace negotiations and that colonial subjects spared the internationalism of the League of Nations, can we truly understand my twenty-first-century interpretation. It is not just the devastating warfare itself and the four-year and four-month military catastrophe; it is also about Compiegne, the Armistice deal that developed into the mortally diverging paths of WWI demobilized survivors such as Frank and Hitler. Instead, Hitler, who went on to become a revanchist and post-WWI revisionist party leader, argued time and again—the 1939 speech in Berlin is most illustrative—that he was delineating what George Kennan would, decades later, call the "Urkatastrophe" (original catastrophe), great seminal catastrophe of the twentieth century whose aftereffects resulted in deadly shockwaves throughout the history that followed. I maintain that 1918-1919 made that happen.

The Aufa100 Transnational Reappraisal Initiative that I have organized as a follow-up to the first centennial seeks to further a better understanding of the post-war Paris/Weimar/Versailles outcomes. We happen to be the only Europeans participating in the third post-centennial edition of the Doughboy Foundation's 2021 Bells of Peace.[3] Despite the silence left intact by the national and European-Union narratives as Germany was coming to terms with the Third Reich's crimes, it is my hope that it will no longer be easy to ignore the central presence of WWI in subsequent world history. As such, we can finally raise intriguing questions such as: Why did the war for 1917 newcomer United States last as long as for Japan, the Europeans and their colonists, i.e., four years and a few months?[4] My connections to the Anne Frank House have persuaded me that the interwar period needs to be more closely reviewed to better understand the horrible event of genocide and how to prevent it. Stay tuned with us looking forward to the hundredth anniversary of Anne's birth.

Peter de Bourgraaf is the founder and director of the Berlin/Amsterdam-based Aufa100 - Transnational Commission for Reappraisal and Commemorative Culture from 1914. The independent historian and lecturer from the Netherlands published German and English-language monographs such as his lifework at the closure of the First World War centennial *Hundert Jahre Urkatastrophe. Der Kolonialvertrag 1919* [in transnational German]. In a handful of countries, his lectures and book presentations on the European centennial's anticlimax and this war's open end were in demand. Between his graduations in Amsterdam and Hamburg, the Dutchman worked in the visitor's services of the Anne Frank House. De Bourgraaf can be reached at bourgraaf@aufa100.com

Notes

¹ Peter De Bourgraaf, *Hundert Jahre Urkatastrophe. Der Kolonialvertrag 1919* [Hundred Years of Seminal Catastrophe. The 1919 Colonial Treaty] (Göttingen: Cuvillier Verlag, 2018) pp. 95, 98, 144.

² George Orwell, "Review of Adolph Hitler's 'Mein Kampf,'" Review of Adolph Hitler's "Mein Kampf" (Project Gutenberg Australia, February 2016), https://gutenberg.net.au/ebooks16/1600051h.html and Daniel Miessler, "Orwell Reviews Mein Kampf and Perfectly Captures Trumpism," Daniel Miessler, February 23, 2021, https://danielmiessler.com/blog/orwell-reviews-mein-kampf-and-perfectly-captures-trumpism/.

³ "Bells of Peace," Doughboy Foundation, accessed January 22, 2022, https://doughboy.org/activities/bells-of-peace.

⁴ Peter de Bourgraaf, "'The First Peace':: Aufa100 - Transnational Commission for Reappraisal & Commemorative Culture from 1914," Aufa100, August 24, 2021, https://www.aufa100.com/l/back2waffenstillstandsinitiatoren/.

part eight

CRITICAL RESPONSES TO WWI LITERATURE

Faulkner Stole My Father's War Wound: Centennial Reflections on William Faulkner and John Reid*

PANTHEA REID, PhD

In 1998, Staige Blackford, then editor of *Virginia Quarterly Review,* adjusted his publication schedule so, eighty years after the Armistice, the autumn issue included my essay, "William Faulkner's 'War Wound:' Reflections on Writing and Doing, Knowing and Remembering." It was about Faulkner's confiscating the record of my father's World War I wound and claiming the story for himself, but my title did not mention my father. Even as I made a case for Faulkner's unacknowledged debt to my father, I failed to acknowledge my own debt. After 2018, Centennial of the Armistice, I revised my 1998 essay to explore my debts to John Reid and to John Fischer, as well as to William Faulkner.

Advertisements in the *Owensboro Daily Messenger* of Friday morning, Nov. 15, 1918, depict a typical world of intractable ailments and unproblematized patriotism. Remedies are promised on most pages for catarrh and biliousness, colds. Articles and ads fretted about the stamina of girls, "tired, nervous mothers," and draftees. A butcher offered rib roast at twenty-five cents a pound as protection against Spanish flu. An ad featured soldiers and Uncle Sam naming a bank "one of the most patriotic." Daviess County, Kentucky, was "near top" in raising war funds.

Among war propaganda, another photo appears on the front page. Its caption reads: "Sergt. John Reid Wounded in France." Beneath the picture of Reid, my father, taken in 1917 when he enlisted, is a letter from American Red Cross Base Hospital No 18, France. It begins: "Dear Mother: I got wounded in the head..." The letter is not as cryptic as standard Field Service Post Cards, which allowed injured servicemen to say they remained "quite well." This letter proved John Reid was alive, though he could not write. Later a newspaper column called "Bugle Calls" further softened Reid's story by suggesting Reid had written "with his left hand." Someone glibly wrote, using Reid's name, "I will be able to use my arm in two or three weeks."

My father's right arm was permanently paralyzed, limp, useless.

FAULKNER AND REID FIRST TRY FLIGHT

By 1914, when World War I broke out in Europe, fantasies about airplanes had existed since the Wright brothers 1903 flight, but neither Central Powers nor Allies expected them to play a military role. In the summer of 1917, the Allies introduced new fighter planes: the British Sopwith Camel and an upscale version of the French Spad, planes contesting the air supremacy of the German Fokker and D-3 Albatross. Air combat became legendary, as Allied and Central pilots eluded and attacked each other with twists, spins, half rolls, vertical reverses, and unexpected zooms from behind cloud banks. Balancing the brilliancy of twentieth-century aerodynamics with the chivalry of the Middle Ages, pilots saluted their enemies before shooting them out of the sky.

Tales of death-defying bravery of flying aces as the American Rickenbacker, the Canadian Bishop, the Frenchman Fonck, and the German von Richthofen, fired the imaginations of John Reid, my father, and of William Falkner, as he then spelled his name. In Kentucky four Reid Brothers and in Mississippi four Faulkner brothers made youthful attempts at flying from trees and rooftops. The average life-expectancy for pilots of biplanes with wooden frames and rotary engines spinning at only ON or OFF was fourteen days. Reid and Faulkner must have known the risks. Apparently, neither cared. Both volunteered for pilot training.

The Army's Air Corps quickly rejected Faulkner. He was under-educated and a drinker. Though only twenty, Faulkner had already discovered the attractions of Memphis, Tennessee, a notorious center for the blues, prostitution, gambling, and easy liquor. Most embarrassing was the official reason: at just over five feet tall, he was too small. Others' successes heightened his humiliation. The Marines accepted his younger, six-foot-tall brother Jack. Also, Faulkner's friend Frank Smythe, a recent graduate of "Ole Miss" medical school, joined the medical military's aviation section and departed for the Mediterranean theater. "Egypt or Italy, his exact destination being unknown," as a Memphis paper tauntingly phrased it.

Along with three brothers and two sisters, John Reid had been raised outside Owensboro, Kentucky, in an idyllic setting—a graceful Victorian home, down an avenue of maple trees, among the orchards his Scottish father Allan had begun planting in 1873. In 1915, John Reid earned a B. S. in the emerging field of electrical engineering and was immediately employed by Western Electric. In 1917, Reid resigned to return to Kentucky and volunteer at Fort Thomas that November for flight training. With a thesis on heat conductivity, a respectable height of five feet, nine inches, and, according to military documents, an "excellent" character, he was ideally suited for officer training. Reid breezed through all paper tests. Then he was strapped into a hammock and twirled about, side over side. Unable to refocus his eyes, Reid could not raise his head, for he was seized by spasm after spasm of outraged vomiting. He flunked, disgracefully, the mandatory air-sickness test.

Like Faulkner, Reid was humiliated. Since he could not be a pilot, he refused to be an officer. Unlike Faulkner, he enlisted as a foot soldier, half expecting to die in battle. He nearly did.

Unlike Reid, Faulkner covered his humiliation by drink. He would escape from Oxford to Memphis, often ending up at the Smythes. He stayed in Frank's empty room and heard whatever his sister, Carolyn Smythe, knew about her brother's service in the medical military. When Faulkner failed to return from Memphis, his mother, "Miss Maud," would call Carolyn asking her to loan Billy train fare home or, after he was found on the Smythes' porch swing in the middle of the night, to buy a ticket and put him on the morning train.

In 1917-18, Dr. Frank David Smythe, father of Frank and Carolyn, was recruiting volunteers for the war effort among, as a Memphis paper put it, "the leading members of the medical profession," including his son Frank.

REID GOES TO WAR, FAULKNER GETS QUARANTINED IN CANADA

In spring of 1918, while John Reid and the quickly assembled American forces trained state-side, the German Chief of Staff, General Ludendorff, received an influx of men and munitions to launch a major advance. Only the landscape, decimated into a muddy quagmire by almost endless trench warfare, seemed capable of slowing German advance.

That spring, the parents of Estelle Oldham, Faulkner's sweetheart, announced her engagement to Cornell Franklin, a Mississippian now enjoying a lucrative law practice in Hawaii. Billy Faulkner had thought he was her fiancé.

Having failed at both war and love, he fled north. In Connecticut, he clerked in a munitions factory, wrote poetry about a figure paralyzed by failure and loss, and tried to enlist in the Royal Air Force.

In his enlistment photograph, John Reid seems confident and imposing. Rushed out of training to stop the Germans, Reid and the Third Battalion arrived in Brest aboard the Agamemnon on April 18. That same day, Estelle Oldham married Mr. Franklin, not Mr. Fualkner.

By late May, American forces were almost ready to engage the Germans. In an early June letter, written atop the lid of his mess kit, John Reid assured his mother: "the Huns got loose about a week ago & made some gains but I hear they have driven them back now & I dont think it will be long till we will all be coming home." He imagined the Reids, who never ate a meal without some sort of fruit, "living on strawberries & cherries now." He promised to get home to the Reid orchards "in time to eat my share of the preserves." In a postscript, John asked her to save a clipping labeled "Happy American Marines riding in 'side-door' Pullmans to the front in France."

The waving Marines were not happy for long. Reid's platoon rode "three nights & two days (forty of us) in one of these Pullmans & they are not nearly as big as a first-class boxcar at home. It sure was a great ride." That irony remains Reid's only hint the Great War was not as glorious as promised.

By then, Dr. Frank David Smythe's medical volunteers were also in France. Smythe, now a major in the U.S. Army Medical Corps, oversaw a hospital at Juilly, about thirty-five miles from Paris. Lt. Frank Smythe left the aviation section to join his father's medical corps.

A member of Company H, Seventh Regiment, Third Division, of the American Expeditionary Forces, Reid soon found the "Huns" were not easily "driven" back. He fought from the beginning of June in the crucial battles of Chateau-Thierry and Belleau Wood, where Jack Faulkner, William's brother, with the Fifth Marine Regiment, also fought.

Given heavy casualties, the British lowered standards for pilot training. Masquerading as a British citizen, Faulkner got himself accepted in the Canadian branch of the Royal Air Force. From mid-summer, 1918, he trained in Toronto. In his cadet uniform, the twenty-year-old looks small and frightened.

His letters home, however, including one recounting the air-sickness test he barely passed, strike a self-assured pose, signaled with a new name— "Faulkner," as he said his Scottish fore-bearers had spelled their name.

With the Third Division, John Reid fought the entire month of July: in the Marne Sector, July 1-14, the Champagne Marne Defensive 15-18 July, and the Aisne-Marne Offensive July 18-29, the turning point of the war. In that offensive, Reid rescued his wounded sergeant under heavy machine-gun fire. Leading a seemingly charmed life, Reid was promoted to sergeant and later received the Silver Star for "gallantry in action at la Theodorie Farm, France, July 22, 1918."

On September 1, John Reid sent home a letter marked "For Papa Only" asking, if that he were killed, his life insurance policy be divided equally among his five nieces and nephews to "go toward their education." Then he, like Jack Faulkner, helped cut off the retreating Germans in the St. Mihiel offensive in mid-September.

Inadequately trained American troops faltered in the Meuse-Argonne offensive of September 26. Then the next day, Jack Faulkner, with the 5th Marine Regiment, helped a French army corps rebuild roads decimated by four years of fighting so more experienced Allied forces could advance. Reid's division trekked by night over those make-shift roads northwest toward the area between the Meuse River and the Argonne Forest, the most vital point in the German defenses: a zone of trenches, barbed wire, field fortifications, and railway supply lines, which was about ten miles deep and presumed to be impregnable.

The Memphis papers carried stories about the probability of a Prohibition amendment and about the gallantry of local heroes, including Smythe's medical volunteers, a pilot named Mack Grider, and Jack Faulkner. William Faulkner remained in Toronto, where the entire city, including the School of Aeronautics, was under quarantine, thanks to the Spanish influenza epidemic.

Reeling from the St. Mihiel defeat, weary of capture and death, fearing a renewed offensive, on October 4, the Germans petitioned President Wilson for an immediate cessation of fighting. Dr. Smythe must have had high-placed connections, for that day the *Memphis News Scimitar* wrote: "SMYTHE THINKS BOCHE WILL QUIT." The "Boche" were indeed ready to quit, but President

Wilson delayed, Germans rearmed, and General Pershing ordered American troops to advance in rain and sleet against reinforced German defenses.

In the advance of October 4 and 5, Sergeant George Gunter oversaw the first platoon and Sergeant Reid commanded the second. Charging machine-gun nests in the dense hilly woods, both platoons met heavy artillery, poisonous gasses, and machine gun fire. Sergeant Reid reformed his lines and reattacked, repeatedly. Then, probably in the Bois des Ogons, John Reid's luck ran out.

A machine-gun bullet cracked his helmet and tore into the left of Reid's skull.

Against orders and under heavy fire, Corporal Leon A. Dombrowski spent twelve hours dragging Reid back behind the lines. Thinking Reid "could live only a few hours," Sergeant Gunter then saw to his transport by stretcher for more than a mile to Cierges. From there, motorized ambulance corps evacuated the wounded to Montfaucon. Gunter kept Reid's watch, expecting to return it to the Reid family, as he later explained to John Reid, "with details of your daring death."

Not having heard for three weeks, on October 18, Allan Reid stopped pruning his youngest trees, to write to John "I hope the Huns have not got you." The letter was returned unanswered, which must have provoked anxiety on the Reid Farm. On the battlefield too, Reid was assumed dead. Sergeant Gunter searched to find Reid's dog tag on a fresh grave so he could send it to Reid's parents.

In a month, Dr. Frank David Smythe took charge of the 1,000-bed base Hospital #57 in Paris. As one of three Chiefs of Surgical Service at the hospital, Smythe was promoted to Lieutenant Colonel. That month, Hospital #57 treated as many as 2,000 patients, possibly including John Reid. On November 1, Jack Faulkner suffered from gassing and shrapnel wounds to the knee and scalp. President Wilson accepted the German surrender. Still in training, William Faulkner wrote his mother ruefully, "It looks like the whole thing is over."

REID UNDERGOES BRAIN SURGERY, FAULKNER FEIGNS A LIMP

After six weeks of silence, a letter posted in France and written in an unknown hand, arrived at the Reid farm. The *Owensboro Daily Messenger* printed it, along with a picture of John Reid and news of the armistice, on the front page. Here is the complete letter:

Dear Mother: I got wounded in the head, and I can't sit up enough to write myself, but will as soon as I can. I am getting on fine, and just wanted to give you my love, and tell you not to worry. Love to rest of family. John.

That news was hardly reassuring. John was alive but too injured to write himself. He suffered, as his honorable discharge papers would read, from a wound in the "parietal bone left, with injury to brain," "hemiplegia, spastic right," and moderate aphasia.

On November 11, William Faulkner was one week short of completing ground school. He had probably not flown and certainly, he had not piloted a plane. Quarantined, he had not even left Toronto.

Reid endured a delicate operation to remove the bullet and debris in his head. With his survival possible, Reid was shipped with the most seriously wounded soldiers to the states aboard the U.S.S. *Antigone,* which left France on December 22, 1918. John Reid's elder brother Robert traveled from Kentucky to meet him at Newport News, Virginia, in early January 1919. When Robert boarded the ship, however, he found so many soldiers encased in bandages, he couldn't tell one from the other. After Robert Reid found his brother, John was taken to Walter Reed Hospital in Washington DC. The *Owensboro Messenger* reported that because of the surgery at Walter Reed, "the paralysis in the right arm and leg can be relieved."

John's fellow soldiers wrote to the Reids, asking if he had survived. One explained, "John was a brave gallant soldier, always out in the worst shell fire, doing everything he could for his comrades"—including loaning the men money. Sergeant Gunter returned his watch and regretted he could only collect seven francs, probably because "most of the fellows who owed you are in the hospital"—or dead.

In early December 1918, William Faulkner arrived in Oxford wearing a second lieutenant's uniform, "wings" for completed pilot training, and a cap signifying overseas service. He leaned on a cane and limped. He claimed he ended a celebratory joy ride by crashing his plane into the hangar, where he hung upside down.

In Tennessee, Colonel Smythe delivered inspiring talks on the heroism of Memphis nurses during the war. Meanwhile, Faulkner's official papers arrived home. Under the heading "Casualties, Wounds, Campaigns, Medals, Clasps, Decorations, Mentions, Etc.," the word "NIL" is stamped. Yet Faulkner paraded about Oxford in a uniform, his limp and overseas insignia inspiring salutes from soldiers. He was photographed in at least six different combinations of military garb, looking jaunty, dapper, and self-assured.

Probably, in early 1919, a soldier stationed at a nearby army camp arranged to meet Carolyn Smythe in the lobby of the old Peabody Hotel. When they met, William Faulkner also appeared "in uniform and bandaged." His arm was in a sling, and he walked with a cane. Faulkner told Carolyn's beau, who later told Carvel Collins, that, when Faulkner's plane was hit, the British Major with him had been killed. Faulkner had also said he had "fallen through trees so that he was only injured, not killed."

Late in 1918, when she, at last, felt her son would live, Marion Reid, herself a formidable Scotswoman, wrote Corporal Dombrowski at the military hospital in Fort Dix, New Jersey, requesting information about her son's wounding and rescue. In January, Leon Dombrowski answered Marion Reid's query. A Polish immigrant with imperfect English, he began: "I am trying to be real frank with you Mrs Reid and will answer the questions as good as I can and so you can understand them."[1] On ten pages of a lined tablet, Dombrowski laboriously detailed the agonizing events of October 5-6, 1918, when he rescued the paralyzed John Reid. For much of 1919, Walter Reed surgeons tried to rehabilitate John Reid's brain and body through surgery. Contrary to newspaper optimism, paralysis was not easily "relieved."

FAULKNER POSSIBLY GETS IDEA OF HEAD WOUND FROM REID

In 1919, William Faulkner wrote static poems about failure and loss. Sometimes he donned the uniform of a soldier who had been overseas. Calvin S. Brown remembered Faulkner told neighborhood boys he escaped injury when he crashed in France because the "thatched roof of a peasant's house broke his fall. In fact, he added, he went right through the roof and landed unhurt in the middle of a big tureen of soup, with the family sitting around it at the Sunday dinner table!"

That fall William Faulkner enrolled as a special student at the University of Mississippi. Though college fraternities and alcohol were both outlawed, he joined the now underground Sigma Alpha Epsilon, whose alumni included Faulkners and Smythes. Among current "brothers" was Ben Wasson, who would later work as Faulkner's agent. Wasson remembered the SAEs enjoyed Faulkner's drunken storytelling but were skeptical about the veracity of his tale of crashing a plane and "receiving a leg injury that caused him to limp."

On December 16, 1919, at the Southern Surgical Association's meeting at the elegant Grunewald (now the Fairmont) Hotel in New Orleans, H. H. Kerr, M.D., a Canadian who had become a Lieutenant Colonel in the U.S. Medical Corps, gave a talk on "The Late Treatment of Gun-Shot Wounds of the Head." Given that Dr. Frank David Smythe was a Lieutenant Colonel, one of three Chiefs of Surgical Service at hospital #57, and a Fellow of the Southern Surgical Association, he must have been interested in, perhaps involved with, Dr. Kerr's talk.[2]

Considering the "defect syndrome" that injured patients exhibited, Dr. Kerr optimistically assumed surgery would cure psychological symptoms, along with most physiological ones. Comparing two methods of treating wounds to the head, either through inserting a metal plate in the skull or through making a bone graft, Kerr explained at Walter Reed, they preferred the latter method, which he illustrated with pictures of his patients' skulls. The New Orleans and Memphis papers both gave cryptic accounts of the surgeons' convention, printed as Dr. Smythe traveled home from New Orleans.

Carolyn Smythe had been a friend of the Memphis ace, Mack Grider, whose daredevil flying record and dramatic death, along with her father's and brother's service in France, were likely topics of conversation whenever Faulkner visited the Smythes. The elder Dr. Smythe seems also to have talked about new surgical methods for treating war wounds to the head. Though the association's journal was on the shelves in the Ole Miss library in Oxford, Smythe probably shared with Faulkner his own copy of the *Transactions of the Southern Surgical Association* which reprinted Dr. Kerr's talk with its accompanying illustrations.

Two of those illustrations are of John Reid's skull, showing his shaved head, first battered-in and then patched together. After the grisly surgery, his head looks pitiful.

I believe Faulkner heard about Dr. Kerr's talk from Dr. Smythe and "got" his wound by reading Dr. Kerr's article. In the early 1920s Faulkner's "war wound" metamorphosed. He no longer suffered a rather vague injury in arm and hip, producing a limp. Suddenly, Faulkner had a specific wound to the skull, supposedly covered by a silver plate.

With his own wound to his left skull covered by a bone graft from his right skull, John Reid remained in Walter Reed until May of 1920. Through relentless physical therapy, he gained partial use of his right leg. In a "graduation" photo, he stands straight among his class, but his good left arm has to hold his limp right one behind him. John Reid grimaces.

When Faulkner arrived in New Orleans in the fall of 1924, he introduced himself to the Bohemian crowd in the French Quarter, including the prominent writer Sherwood Anderson, as a wounded veteran. Anderson memorialized Faulkner in "A Meeting South," a story about a little southern poet, formerly an aviator who has survived a crash in the Great War. The poet drinks prodigious quantities of illegal liquor to ease the pain caused by a silver plate covering a wound in his head. Anderson's poet passes out on the brick patio of a bordello talking to the madam, a New Orleanian called "Aunt Rose" Arnold.

When "Miss Maud" read Anderson's story, her son did some fast explaining. He maintained Anderson made up the tale about him because truth "never makes a good yarn." Certainly, Faulkner's own non-experience of the Great War did not. But the RAF uniform with the overseas service cap, the crash, the limp, the head wound, and the silver plate in his skull made a very good yarn. In New Orleans he lived the life "of a war hero."

In New Orleans during the first half of 1925, Faulkner began his metamorphosis from failed poet into great novelist by writing *Soldiers' Pay*. This first novel offers another suggestion that he had indeed read Dr. Kerr's article. Donald Mahon's symptoms from a gunshot wound to the head resemble the "defect syndrome" Dr. Kerr described. Mrs. Powers' faith in surgeons almost parodies Dr. Kerr's self-confidence. Mahon soon dies from his wound, but unscarred Cadet Lowe envies the maimed Lt. Mahon. While Lowe desires to swap his healthy body for Mahon's shattered skull, the healthy William Faulkner was masquerading as John Reid: a wounded veteran with a shattered skull.

Contrary to Dr. Kerr's optimistic report on the newest procedures of 1919, surgical patching partially mended the hole in John Reid's skull but did not address the psychological aftermath of losing a possible fiancée, a promising profession, agility in the right leg, and use of his right arm. Nor could it address the lifetime "defect syndrome" Reid suffered. In 1927, John Reid suffered a terrible relapse. His brothers, Robert and Allan, Jr., rushed him back to Walter Reed. As they maintained twenty-four-hour watches, his wound was reopened, his surgery redone, for the final time.

FAULKNER IMMORTALIZES WOUND IN WRITING, REID REMAINS A QUIET HERO

William Faulkner was writing stories and his third novel. It was about twin brothers, both dare-devil aviators in the Great War. Faulkner's John Sartoris flies his Sopwith Camel straight up into cloud banks, where the more air-worthy German Fokkers surround him, using tactics Baron von Richthofen had refined. When his plane catches fire, as his brother's does not, John jumps without a parachute. As he goes down, John Sartoris's last chivalric gesture is to thumb his nose at his twin brother, Bayard.

Bayard Sartoris lives but is haunted by guilt because he has played the game of war more cautiously than his twin. After returning to Mississippi, Bayard engages on one foolhardy gesture after another until he finally kills himself in his own airplane crash. Like the dynamic between Mahon and Lowe in *Soldiers' Pay*, the fictional Sartoris twins' story seems a parable for Reid's wound and Faulkner's guilt over neither fighting nor being wounded. He expressed that guilt in the words of another Bayard: "those who can, do, those who cannot and suffer enough because they can't, write about it."

In 1929, when he was finishing *The Sound and the Fury*, Faulkner married the once-lost, now-divorced, and at-last-available Estelle. In 1938 Reid married Nell Marshall, an Alabamian, who held an MA in mathematics from Columbia University but worked during the Depression as a YWCA staffer first in Owensboro, Kentucky. Each marriage brought both happiness and bitterness. In each, bitterness was alleviated by the birth of a single daughter, born late to the Faulkners.

I was born very late to the Reids.

When *The Sound and the Fury* was published, in September 1929, Faulkner surreptitiously took flying lessons at the Memphis airport. He cashed in on an aviation fiction vogue by writing stories, based supposedly on his own, but in reality, on Mack Grider's flying exploits. As if to pay his debt, Faulkner wrote a tribute to Grider for the *Memphis Commercial Appeal.*

In 1933 Faulkner began secretive, prolonged, and at last successful flying lessons. Having written about the seamy side of Memphis in the notorious *Sanctuary,* however, Faulkner had lost his anonymity. He deflected reporters' curiosity with another fabrication, explained in late March by the *Commercial Appeal*'s headline "Recovery of Lost nerve is Author's Aim." The subtitle was "Faulkner, War Flier, Seeks to Use His Wings Once More."

By the 1940s, Faulkner tried to expunge "war flier" stories from his biography, but when his nephew became an aviation cadet, Faulkner reverted to fantasy: "I would have liked for you to have had my dog-tag, RAF, but I lost it in Europe, in Germany. I think the Gestapo has it; I am very likely on their records right now as a dead British flying officer-spy." Faulkner's fame had not only aroused reporters' curiosity about his flying lessons, but his own brothers also imitated him by adding a "u" to their "Falkner," making them too "Faulkner."

In 1948, after we moved to Alabama, John Reid became a secondary figure in his mother-in-law's house. We spent all vacations back on the bucolic Reid orchards until my father's beloved Victorian home-place burnt down, and he lost most of his connections with his agrarian past. As a teen, I was embarrassed by my father's difficulty with simple tasks, like writing with his left hand or climbing stairs. No handicap ramps existed, so he laboriously climbed the church steps using his left hand to hold a railing and dragging his right foot behind, furious if anyone offered to help. He mastered driving by tediously shifting the "fluid drive" gears with one leg and steering with one hand.

With a wound in the left brain where the nerve-center for speech is located, John's moderate aphasia also escalated. A photograph of my father carried an official star. When I learned it was the Silver Star, I begged to hear of his heroic exploits. He remarked he had rescued a man who died anyway. His only confession about risk-taking in the Great War was that he had smoked Camel cigarettes.

When I desperately needed adult counsel, I found my female relatives too conventional and my father unable to put thoughts and feelings into words. He disapproved of my boyfriend but could not explain his objections and worries. I married the guy.

FAULKNER: REID'S "DARK TWIN"

After winning the Nobel Prize, William Faulkner became a public spokesman for equality, tolerance, moderation, privacy, the efficacy of art, and the "old verities." By 1962, when Faulkner died, my father had all but ceased to read and to speak, except in monosyllables. By the 1970s he had lost most of his ability to talk. In 1974, I left my teaching post at Virginia Tech and my first husband for a post at LSU. There I met and fell in love with the English Department's eighteenth-century scholar, John Fischer. After I brought John to Alabama to meet my parents, he read about medical treatments for aphasia, hoping to understand and help my father, but it was half a century too late. Still, Daddy reached out with his left hand and clasped John's to him, clearly grateful I now had a good man in my life. By 1980, when Daddy died, he was unable to make such gestures. We buried what seemed an empty husk, though he had been (to use a Faulknerian term) the "dark twin" of William Faulkner.

Once, when he heard me laugh over Faulkner's war-wound fabrication, John Fischer said, "That's your father's story, as you know." But, as I admitted to John, I hadn't let myself know. I had longed for a more dazzling father than the crippled old man a machine-gun bullet left me, who could or would not talk about his own bravery.

After my mother died in 1990, among box after box of worthless papers, I discovered Dr. Kerr's article. Reading the article with its photos of my father and finding Dr. Kerr had given the talk at a 1919 New Orleans convention of surgeons among which Dr. Smyth was a fellow, I knew Faulkner must have read it too. Among my inheritance of books and letters, maps, clippings, and military records, my most dramatic discoveries were articulate pre-war letters from my father, written in a graceful handwriting I had never seen, and Leon Dombrowski's account of the events of October 5-6, 1918. Reading that testimony was as powerful an experience for me as studying the commissary records is for Faulkner's Isaac McCaslin. While Isaac discovers such shame and perfidy in his family that he revokes his heritage, I discovered such bravery. I was only ashamed of my own blindness.

Even with misspellings and almost non-existent punctuation, Dombrowski's narrative achieves an immediacy that rivals Faulkner's famously unpunctuated stream-of-consciousness passages. (The entire letter, with my commentary, *VQR* 74, 4 (611-16) is now online.)

> . . . I was about 15 yards away from John with my squad of men seeing John fall over I said to my self I must reach Him after crawling from shell hole to shell hole I found out that John was Kill'd at this time I was about 5 yards from where Sgt John lay, the Boys saying that he was hit in the Heart I yelled over "John John" but no answer. so I gave [up] all hope of seeing John alive. Laying there about another hour I heard John gave a moan which I heard and said, to the rest of the Boys "he is alive!".
>
> . . . I have told Gunter about John and that sure did hit Sgt Gunter very hard. there we were dark, and John wounded so seriously so I told Gunter to see if something could be done.

Sergeant Gunter later wrote Allan Reid, "After I got [John] off the field I thought he could have only a few hours [to live]. It was a great loss to me there being no other man in the company so close a friend to me as John, and I assure you sir the company never lost a better soldier than he." Leon Dombrowski's narrative continued with some bitter comments about the commissioned officers:

> As far as I know all the Corporals & Sergeants were on the job, but as to the Officers I can say this much. that there was not a bit of fighting spirit in them, what so ever, and I think John can approve of that, that same morning we found that our Company went over the top with out an Officer and we were without them till the day when we received news about being released then they started a comming in we had two Officers at that time 1st & 2nd Lieutenants. of course I was lucky enough to linger 10 or 12 days longer [until he too was wounded] I also saw Sgt Gunter get wounded upon another front to which we went after being relieved only for a day and night.

Dombrowski ended his narration with one final poignant sentence: "All this happened in the bloody Front of Argonne Woods."

When I finished reading that 1919 letter, written on cheap, yellowed note paper, I was crying. My tears were shed for the bravery of Reid and Dombrowski and for the eloquence of the latter, who had given unpolished but passionately expressive language to the former.

Faulkner expanded his narrative experiments and his empathy for an ever-widening cast of characters, but when his own daughter once appealed to him to stop drinking by saying "think of me," he replied: "Nobody remembers Shakespeare's children." John Reid, who spent so much of his thwarted life taxiing me slowly to music and dance lessons, Girl Scouts, and horse-back riding expeditions, would never have spoken or thought so dismissively. Though a machine-gun bullet lost me a more communicative father, I now realize, as it freed Reid from codes of competition and aggression, I was also its beneficiary. Though he hardly talked, John Reid did show, in Faulkner's words from *Absalom, Absalom!* an "overpass to love."

In the sections of *Absalom* set in 1910, Shreve McCannon, a Canadian and future M.D., like Dr. Kerr, asks Quentin Compson to "tell about the South." Together they transmute second-hand hypotheses and a Civil War letter written in stove polish into the story of the American South. Their joint fabrication provides them an "overpass to love."

Uncovering in attics and archives much of the above information, including Faulkner's heretofore unnoted connections with the Smythe family, afforded me an intellectual and an emotional "overpass" into Faulkner's past and into my own.

If the story of my father's literal paralysis helped rescue Faulkner from linguistic paralysis, then John Reid, or the image of Reid and his fellow wounded soldiers, played an enabling role in Faulkner's transformation from a writer of stilted prose and frozen verse into an inventor of kaleidoscopic, dynamic narrative experiments, both energized and energizing.

WWI language displays a startling disparity between rhetoric of glory and often suppressed rhetoric of suffering. misery, and death. That's the disparity between Faulkner's early grandiose posing and my father's humble suffering. This disparity between the "abstract and the actual," as I termed it in my first book, became central to Faulkner's writing.

No doubt, Faulkner and Reid met in my psyche because Faulkner was the glamorous, articulate father I lacked, but he was the self-absorbed, lying, chauvinistic father I also lacked. My father was not a great author, but he was a better father than Faulkner, as John Fischer was to our daughter Hannah.

In the year 2000, John Fischer presented at the Fourth Münster Symposium on Jonathan Swift, an essay on Swift's "Cadenus and Vanessa." He ended his erudite conclusion: "I have also written this essay, itself about books and love, for Panthea Reid."

Now I'd like to dedicate this essay to John Irwin Fischer, who possessed the courage and decency of John Reid and possessed much of the brilliance of William Faulkner. In Leon Dombrowski's words, "all this happened."

*A version of this article entitled "William Faulkner's 'War Wound': Reflections on Writing and Doing, Knowing and Remembering" was first published in the 1998 Autumn Issue of *Virginia Quarterly Review*. Reprinted with permission of the author.

Panthea Reid has edited books on Walker Percy, Ellen Douglas, and (finishing for John Fischer, her late husband) Jonathan Swift. She has authored: *William Faulkner: The Abstract and the Actual* (LSU); *Art and Affection: A Life of Virginia Woolf* (OUP); *Tillie Olsen: One Woman: Many Riddles* (Rutger's), and *Body and Soul: A Memoir of Love, Loss, and Healing* (Wild River Books). She had been a Fulbright Teaching Fellow, an NEH Fellow, and president of the South Central MLA. She holds a PhD from UNC-Chapel Hill and a recent MFA from Randolph College. She is Emerita Professor of English at Louisiana State University.

Notes

[1] The full text of Leon Dombrowski's letter can be found in Reid's 1998 article, "William Faulkner's 'War Wound': Reflections on Writing and Doing, Knowing and Remembering" | VQR Online, accessed January 24, 2022, https://www.vqronline.org/essay/william-faulkner%E2%80%99s-war-wound-reflections-writing-and-doing-knowing-and-remembering.

[2] The elder Smythe probably knew John Reid's story. Army personnel records from WWI were lost in a fire, but I know Reid was taken to a hospital at Montfaucon which treated contagious diseases. Because the mortality rate for operations on head wounds had been as high as 65 percent, by October 1918, the Army ceased conducting hazardous partial operations at field hospitals but endeavored to transfer patients with head wounds to the most experienced neurosurgeons at surgical hospitals, such as #57, where Dr. Smythe was stationed. He might have operated on Reid. Certainly, he would have taken interest in Kerr's lecture and illustrations. He might have passed on the story to a hung-over house guest named William Falkner.

Waking Up to History:
John Dos Passos, the Cut-up, and World War I

M.C. ARMSTRONG, PhD

Ten years before traveling to Iraq as a war reporter, I took an undergraduate class on WWI and the Modernists at James Madison University. It was 1999 and I was an acolyte of the Beats, a history major who suddenly believed his time would be better spent as an English major exploring the virtues of spontaneous prose and the open road. But my professor, Mark Facknitz, saved my soul. If I could not see the dialectic between English and history prior to this course, I certainly recognized it afterward. Facknitz wore a suit and tie to class every day. He was famously sadistic in his demands of his students and would sometimes erupt in outrage when you could not keep up with the reading. And no author seemed more worthy of his passionate advocacy than John Dos Passos. Facknitz structured his course around The Library of America's edition of Dos Passos' *USA Trilogy*. At well over 1,200 pages, *USA* is an ambitious undertaking by virtue of sheer bulk, to say nothing of the density of its prose and its acrobatic use of narrative device. With seven other texts on the syllabus, students dropped Facknitz's course in the first week, but I was mesmerized by both Dos Passos and Facknitz and the way they seemed to speak directly to my confusion as to whether I should study history or English. Facknitz was no neoliberal. He did not believe that the fall of the Berlin Wall and the end of the Cold War constituted an end to history. He seemed to teach with a prophet's conviction that history was merely hibernating, as it always does, between wars.

When the towers fell on September 11, 2001, and the literary world returned to history with what I saw as a sudden commitment to non-fiction, I felt prepared to read the new rhetoric thanks to Facknitz and Dos Passos. If our current wars bear a "phantasmatic" quality, as Marc Redfield suggests in *The Rhetoric of Terror*, Dos Passos offers us a decoder ring for the phantoms, a critical lens focused not on bombs, torture, drones, casualty counts, or any particular topic or theme but on language. Language as history. In his introduction to the 1932 edition of *Three Soldiers,* Dos Passos reflected on the condition of the post-war world in 1919, the year he began to write what was to be his first major novel:

> Lenin was alive, the Seattle general strike had seemed the beginning of the flood instead of the beginning of the ebb, Americans in Paris were groggy with theatre and painting; Picasso was to rebuild the eye, Stravinski was cramming the Russian steppes into our ears, currents of energy seemed breaking out everywhere as young guys climbed out of their uniforms, imperial America was all shiny with the new idea of Ritz, in every direction the countries of the world stretched out starving and angry, ready for anything turbulent and new.[1]

Thirteen years after the completion of *Three Soldiers,* Dos Passos is hardened, disillusioned, and amid the Great Depression. Although he briefly laments the loss of the world of 1919, the purpose of his new introductory essay is not to express regret so much as to assert a theory of writing based on lessons learned. By 1932, Dos Passos had grown into a passionate and well-informed critic of capitalism who viewed the novelist as the producer of a "commodity that fulfills a certain need."[2] Thus, his essay is concerned with how a writer can transcend the base function of "the daydream machine" and, thereby, participate in the shaping of human history. By dealing with the speech of his generation and making "aspects of that speech enduring by putting them into print," the writer's function expands to that of "the architect of history."[3] For artists caught up in the relatively new world of mass-produced media, Dos Passos believed that it was imperative to "deal with the raw structure of history now . . . before it stamps us out."[4] This was in the introduction to Dos Passos' *Three Soldiers* that hooked me on the impact of WWI on the Modernists.

When my generation went to war under the banner of a battle against an "axis of evil," I was not shocked or awed. Dos Passos' "raw structure of history" is a map of speech in which we can witness the virality of language, phrases like "axis power" mutating into "axis of evil"; "The Cold War" and "the war on drugs" morphing into "the war on terror." The "raw structure of history," according to Dos Passos, is made of protean language, an increasingly opaque linguistic collusion between commercial and national forces, a blurring of the line between the private sector and the public sphere slash square. This commerce is what historians must study.

The onset of WWI, as Raymond Williams describes it, forever altered the relationship between language and power. Williams describes the advertising industry prior to 1914 as crude and "quack," and although occasionally manipulative, insofar as it played on people's fears of illness, transparent. However, with WWI came the "psychological warfare" from agencies such as George Creel's Committee on Public Information. Williams' primary example is informative:

> Where the badly drawn men with their port and gaspers belong to an
> old world, such a poster as 'Daddy, what did YOU do in the Great War?'
> belongs to the new. The drawing is careful and detailed: the curtains,
> the armchair, the grim numb face of the father, the little girl on his knee
> pointing to her open picture-book, the boy at his feet intent on his
> toy-soldiers. Alongside the traditional appeals to patriotism lay this kind
> of entry into basic personal relationships and anxieties.[5]

The techniques that public information bureaus developed in WWI were taken into the marketplace when the war ended. WWI, Dos Passos said once in an interview:

> Was my first experience with the fantastic way people's minds become
> imprinted with slogans. Overnight, almost, men I'd known at Harvard
> who were quite respected—I won't mention their names—turned from
> extremely reasonable beings into fanatical Hun haters.[6]

This moment of crisis in language, although certainly not the first instance of censorship or state-sponsored media, is one of the central events of Modernism. From this moment on language takes on an overt appearance of threat and, thus, requires confrontation.

For Dos Passos, and later for writers such as William S. Burroughs and George Packer, the cut-up is the historian's way of waking the reader up to history. The cut-up is a central device in the *U.S.A.* trilogy. The "Newsreel" sections, in which Dos Passos' splicing and merging of phrases challenges the reader's reliance on associative blocks of language, is perhaps one of the first places where we see cut-up overtly incorporated in popular American literature. In these sections, the reader sees the artifice, the construction. *U.S.A.* is full of instances of theory becoming practice, characters engaged in cut-up. Dos Passos lets the reader experience the experiment as both guinea pig and scientist. It all begins with an ad. In the first pages of *U.S.A.* Dos Passos sets up a major theme of the trilogy: the word for sale. Mac McCreary, the first character, gets his first job working for his uncle Tim's printing press: "The first print Uncle Tim set up on the new machine was the phrase: Workers of the world unite, you have nothing to lose but your chains."[8] When his uncle's shop is closed by a bank because of his pro-labor stance, Mac goes to the classified ads to find a new job. The first advertisement that catches his eye reads: "Bright boy wanted with amb. and lit. taste, knowledge of print and pub. business. Conf. sales and distrib. Proposition $15 a week apply by letter P.O. Box 1256b."[9] The advertisement is a con. What the advertiser really wants is an errand boy to help him sell books around the country, but he's not interested in paying fifteen bucks a week. Mac bites on the bait. He never sees a cent. The novel thus begins with a small-scale con by a corrupt peddler named Doc Bingham that foreshadows the larger cons of J. Ward Moorehouse, the public relations executive whose involvement in the CPI renders him the central figure of the *U.S.A.* trilogy.

U.S.A. is a tragic story about the growing disillusionment of characters such as Mac who constantly fall prey to lies, the poetry of the state and the syndicates. Dos Passos' use of cut-up is often ignored by critics, but it's fundamental to his critique of capitalism and American democracy. Through cutting up the famous slogans of his time, such as Woodrow Wilson's promise to "make the world safe for democracy," Dos Passos threatened to turn the words of politicians against themselves, to revolutionize language. In "Meester Veelson," Dos Passos sketches a brief biography of Wilson, setting his life in the context of one who grew up "in a universe of words linked into an incontrovertible firmament by two centuries of Calvinist divines."[10] As the son of a Presbyterian minister, "God was the word/ and the word was God" for the future president.[11] Years later, during the war, this mentality would influence Wilson's policies concerning the media. Throughout the "Meester Veelson" chapter, Dos Passos quotes the president at length in order to demonstrate Wilson's belief in rhetoric and its consequences, showing his speeches to be characterized by poetic flourishes and direct commentaries on the importance of harnessing the powers of the press: "We are witnessing a renaissance of public spirit," Wilson said in his first term, "a reawakening of sober public opinion, a revival of the power of the people, the beginning of an age of thoughtful reconstruction."[12]

Using cut-up, Dos Passos holds Wilson to account on the slogans that he used to get elected, such as "He kept us out of war," and the slogans he used to promote the war: "If you objected to making the world safe for cost plus democracy you went to jail with Debs," Dos Passos writes, demonstrating the power of cutting into a slogan and confronting "the raw structure of history," "the speech of our time."[13] The effect of blurring the lines of these texts renders them interactive and desanctified, dialogic where once, during the war, the relationship was monologic and, thus, controlled. In this sense, Dos Passos's novel serves as an eerie precursor for the mashups and contextual violence of the Internet.

The controlling force behind the single voice of the Wilsonian slogans, the man behind the curtain, was, arguably, George Creel. The character of J. Ward Moorehouse, Dos Passos said in a 1962 interview, was largely based on one of Creel's right-hand men, a publicity agent by the name of Ivy Lee.[14] Lee, like the character of Moorehouse, joined the CPI in 1917. He met Dos Passos in a hotel in Moscow as the author was beginning his work.[15] Dos Passos' interviews with Lee led to the creation of a character who seemed to think and speak in cut-up, what we might now call "talking points." Moorehouse, like Mac, embodies the theme of "The word for sale," the individual who has lost his agency and doesn't even know it.

As a young boy growing up in Delaware, Moorehouse wanted to be a songwriter. He showed noticeable talent in the area, but upon reaching adolescence he channeled his skills toward real estate and advertising with a company in Ocean City, Maryland. From there, his fortunes lead him into one of the many horrible marriages of *U.S.A.* and, upon divorce, the city of Pittsburgh, one of the centers of labor controversy in the country at that time. In Pittsburgh, as a journalist, Moorehouse witnesses the relationship between capital and labor and begins to develop ideas for shaping the tarnished public image of both big businesses and the working class. However, when the opportunity arises, Moorehouse drops his low-paying job as a journalist and begins working again in advertising for a man named McGill. When WWI begins, Moorehouse has his own firm, and like Ivy Lee, is viewed as one of the top publicity agents in the country. Like so many men with talent in language, J. Ward Moorehouse joined that historical convergence of corporation and state known as the CPI.

Dos Passos is a careful critic and prophet of fascism. He does not characterize J. Ward Moorehouse as an evil man, but as a character who slowly trades his relationship with capital and labor for one exclusively with capital. Moore-house, quite simply, follows the money and, thus, comes to represent the great Jeffersonian threat to democracy: the faction: the corporation. Moorehouse, like Mac, is often described as thinking in factionalized language, popular phrases, and slogans, dangerous clusters. Talking points. In *The 42nd Parallel*, the first install-ment of the trilogy, we find him on a train to Chicago dictating to one of his many stenographers, Miss Rosenthal:

> He forgot everything in his own words . . . American industry like a steamengine, like a highpower locomotive on a great express train charging through the night of old individualistic methods . . . What does a steamengine require? Cooperation, coordination of the inventor's brain, the promoter's brain that made the development of these highpower products possible . . . Coordination of capital, the storedup energy of the race in the form of credit intelligently directed . . . labor, the prosperous contented American working man to whom the unprecedented possibilities of capital collected in great corporations had given the full dinnerpail, cheap motor transport, insurance, short working hours ... a measure of comfort and prosperity unequaled before or since in the tragic procession of recorded history or in the known regions of the habitable globe.
> But he had to stop dictating because he found he'd lost his voice.[16]

This is the slogan at work, bullet-pointing through the brain on a fast train. "He forgot everything in his own words."[17] He is in a boxcar attached to other boxcars and he is thinking in groupings for groups, for the industry as a whole, factionalization here embodied as a mode of both economy and consciousness. To J. Ward Moorehouse the working class is the recipient, not the cause, of wealth in America. Dos Passos' description and spacing of Moorehouse's thoughts in this segment suggest that his vision is limited by association blocks, which is to say, blocked. As this hyper-efficient corporatized Twitter-like way of thinking leaks from industry into government, these association blocks, these slogans, are what begin to lead the young American men of Dos Passos' generation not just towards products and services but "service" itself, straight into the hate and the bigotry, straight into the trenches and the rapid-fire clusters of the machine-gun.

In 1919, the second installment of the trilogy, the Newsreel shouts: "BONDS BUY BULLETS BUY BONDS" and "AGITATORS CAN'T GET AMERI-CAN PASSPORTS," signifying CPI propaganda and hostility to Americans who might just be pacifists, socialists, or laborers.[18] As Mock and Larson detail, "scholars will long discuss the precise division of "real opinion" in America when war was declared, but there can be no uncertainty regarding articulate opinion as it was expressed in newspapers, books, pamphlets, cartoons, and public address-es—it was overwhelmingly and wholeheartedly on the side of the Allies and in favor of our belligerence."[19] The wide-scale propaganda of the CPI obscured the opinions of "agitators" during the war, using the classic excuse of a state of emergency, the same excuse Hitler and Goebbels used so many times to suppress dissidents. As the character Jerry Burnham says, "a newspaperman had been little better than a skunk before the war but now there wasn't anything low enough you could call him."[20]

The media's collusion with the CPI and the consequent abnegation of its duties as a free press weakened public faith and incited the paranoia that oc-casioned the multimedia techniques of *U.S.A.*, the novel as a "counter-narra-tive" to the nation it critiqued. The cut-up can be seen as a critical response to conventional formats of information that had been suddenly rendered not just inadequate, but lethal. Viral. The challenge for the post-war writer was now quite serious: How to combat a faction of interests who sell death to the poor and uninformed through bigotry and the glorification of war? How to save lives through language? These questions, first introduced to me in 1999 by John Dos Passos and Mark Facknitz, are far from academic and provide the tools allowing us to write the long history of viral language–of how it has been created and how it has been resisted.

M.C. Armstrong is the author of *The Mysteries of Haditha*, published in 2020 by Potomac Books. *The Brooklyn Rail* called it one of the "Best Books of 2020," and the story was nominated for "Best Memoir" at the 2021 Ameri-can Book Festival. Armstrong, who grew up in Winchester, Virginia, embedded with Joint Special Operations Forces in Al Anbar Province, Iraq, in 2008. He published on the Iraq war through the *Winchester Star*. Armstrong has won a Puschcart Prize and his fiction and nonfiction have appeared in *Esquire*, the *Missouri Review*, the *Gettysburg Review, Mayday, Monkeybicycle, Wrath-Bear-ing Tree, Epiphany, War, Literature, and the Arts,* and the *Literary Review*. His first novel, *American Delphi,* will be published by Family of Light Books in 2022. He teaches writing at the UNC Greensboro and is the guitarist and lead singer-songwriter for Viva la Muerte, an original rock and roll band. You can follow him on Twitter @mcarmystrong.

Notes

[1] John Dos Passos, *Three Soldiers* (New York: The Modern Library, 1932), v.

[2] Dos Passos, *Three Soldiers*, v.

[3] Dos Passos, viii.

[4] Dos Passos, ix.

[5] Raymond Williams, "Advertising: The Magic System" in *The Cultural Studies Reader*, ed. Simon During (New York: Routledge, 1999), 418.

[6] Dos Passos, *U.S.A.*, (New York: The Library of American, 1996), 27.

[7] This is one of the primary structural differences between *U.S.A.* and its Beat/Postmodern spawn, William S. Burrough's *Nova Trilogy*. With the former the cut-up is more often described as an effect of characters' interaction with media. With the latter it is the form itself, no formal isolation—no hand-holding—allowed. In other words, whereas Burroughs seems to be conducting an immersive experiment in altered linguistic consciousness, Dos Passos continually zooms in and out on the experiment like a camera eye, allowing the reader to experience both subject and object.

[8] Dos Passos, *U.S.A.*, 25.

[9] Dos Passos, *U.S.A.*, 27.

[10] Dos Passos, *U.S.A.*, 564.

[11] Dos Passos, *U.S.A.*, 564.

[12] Dos Passos, 567.

[13] Dos Passos, *U.S.A.*, 564-567 and *Three Soldiers*, ix.

[14] Lee, the uncle of William S. Burroughs, was "the No. 1 public-relations adviser of American businessmen" during the war (see Pizer, 245).

[15] John Dos Passos, "Interview," in *John Dos Passos: The Major Nonfictional Prose*, ed. Donald Pizer (Detroit: Wayne State University Press, 1988), 245.

[16] Dos Passos, *U.S.A.*, 234.

[17] Dos Passos, 234.

[18] Dos Passos, 445.

[19] James R. Mock and Cedric Larson, *Words that Won the War* (New York: Russell & Russell, 1968), 8.

[20] Dos Passos, *U.S.A.*, 544.

Robert Frost: A Poet for Whom Life and War Were Trials by Existence

JIM DUBINSKY, PhD, veteran

When scholars write of war poets, few consider Robert Frost. Certainly, if the definition of a war poet is one who has experienced the turmoil and vicissitudes of combat, Frost does not qualify. However, if one is willing to consider poets who offer insight into connections between war and the human condition, then Frost surely fits the proverbial bill.[1]

Reading Frost's poems, essays, letters, and gaining insight from a range of biographical perspectives has led me to understand a key component of his personal philosophy: Robert Frost believed in the inevitability of violence. For him, violence and war were natural. He often made statements reflecting this belief. In his private letters to his friend Louis Untermeyer, he argued that "life is like battle" and "war is the natural state of man."[2]

Frost expressed these ideas in response to what he experienced, personally as a husband, father, friend, and artist, as well as what he saw as an individual who lived through both World Wars, the Korean conflict, and the Cold War. They impacted how he understood and framed decisions those close to him made, to include his son's suicide, as I will explain later.

Named after General Robert E. Lee by his father, a Confederate sympathizer, Frost focused on ideals of gallantry and courage early in his life. In early poems such as, "Trial by Existence," Frost emphasizes the importance of valor, both on earth and "in paradise:"

> Even the bravest that are slain
>   Shall not dissemble their surprise
> On waking to find valor reign
>   Even as on earth, in paradise. [3]

This poem was collected in his first volume, "A Boy's Will," which was published, along with his second volume, *North of Boston*, while he and his family were "living under thatch" in England for two and a half years before WWI broke out.

Even though Frost took his family back to the United States in early 1915, he remained engaged with the war and its issues, particularly through his friendship with the English poet, Edward Thomas, who was killed at the Battle of Arras in 1917.

In *North of Boston,* Frost includes several poems that address his perspectives on both the human condition and humankind's proclivity for conflict. Perhaps the most famous statement on human nature comes from "Mending Wall," a pastoral narrative focusing on a narrator who describes an encounter with a neighbor while out walking his fields.

The poem's opening lines speak to the violence that attends the change of season from winter to spring. Most poets focus on the beauty and rebirth that accompany this transition. Frost chose a different path. He opens the poem focusing on how the forces of nature work to tear down walls:

> Something there is that doesn't love a wall,
> That sends the frozen-ground-swell under it,
> And spills the upper boulders in the sun;[4]

The poet shifts focus to signs of spring among humans: hunters and their yelping dogs. The picture Frost paints is one rich with conflict, and the focal point of that conflict is the wall itself, an unnatural but seemingly necessary boundary between neighbors, who are clearly different: "He is all pine and I am apple orchard."[5]

That difference between the neighbors leads the narrator to imagine a potential result stemming from the difference, an image that is at the heart of the poem: the neighbor as an "old-stone savage armed," who "moves in darkness." [6]

This poem, by examining boundaries between neighbors, even those whose only difference is the type of trees on their land (pine vs. apple), delineates an essential conundrum that can be extended to tribes or nations. While on one hand, we often wonder why walls are needed, or as the narrator says, "Before I built a wall I'd ask to know / What I was walling in or walling out, / And to whom I was like to give offence," in the end, particularly after imagining his neighbor as a "savage armed," the narrator concludes by explaining why walls are essential: "Good fences make good neighbors."[7]

While "Mending Wall" might be seen as a reflection on the differences between people and therefore nations that occasionally lead to conflict, poems such as "The Black Cottage," also included in *North of Boston,* focus on the pain and loss caused by war, particularly for those who are not the combatants. Similar to "The Ruined Cottage" by Wordsworth, both poems tell stories of women whose family members go off to war and do not return:

> Said Margaret, 'for I knew it was his hand
> That placed it there, and on that very day
> By one, a stranger, from my husband sent,
> The tidings came that he had joined a troop
> Of soldiers going to a distant land.
> He left me thus—Poor Man! he had not heart
> To take a farewell of me, and he feared
> That I should follow with my babes, and sink
> Beneath the misery of a soldier's life.'
> This tale did Margaret tell with many tears:[8]

In "The Black Cottage," the focus is on the American Civil War, and the story is about one who "fell at Gettysburg or Fredericksburg."[9] The bleakness of the cottage highlights the emptiness of the lives of those left behind, those who strive to make sense of the loss they experience, often resorting to pondering the causes that led to the conflict and wishing they were in a "desert land" with nothing to "covet" or "think it worth / The pains of conquering to force change on."[10]

While Frost is not a traditional war poet, he thought and wrote about war, its causes, and its costs, costs Frost felt deeply and tried to articulate in poems such as "Not to Keep," first published in the *Yale Review* in 1917 and later in *New Hampshire.* This poem focuses on the perspective of a young wife who experiences the joy of having her wounded husband return home to heal, as well as the deep, existential fear of knowing that his stay is temporary.

> . . . The same
> Grim giving to do over for them both.
> She dared no more than ask him with her eyes
> How was it with him for a second trial.
> And with his eyes he asked her not to ask.
> They had given him back to her, but not to keep.[11]

Frost's poems offer insights not only on the personal but also on the public. In "A Soldier," one of several poems in honor of his friend Edward Thomas (another "To E.T."), Frost creates what is almost a hymn to the anonymous soldier who is portrayed as a ". . . fallen lance that lies as hurled/That lies unlifted now, come dew, come rust."[12] The soldier as lance, an image out of place in the modern technological warfare of WWI, hearkens back to Frost's reflections on valor. In this poem, Frost recognizes both the anonymity of the individual soldier, who died by the hundreds of thousands in World War I, as well as the overall uselessness and lack of utility of those deaths with the focus on "unlifted" and "rust."[13]

However, perhaps because "A Soldier" is published in 1928 in the volume *West-Running Brook,* over a decade after "Not to Keep" first appeared, Frost is able to transcend the personal, taking us from the loss of life to the life of the spirit. The poem, a sonnet, concludes with the immediate discomfort we face as these lances descend to earth and make us "cringe for metal-point on stone."[14] The poem encourages us see not only how an "obstacle . . . checked / And tripped the body," but also to see how it "shot the spirit / Further than target ever showed or shone."[15]

In "A Soldier," as he does so often in his poetry, Frost offers a duality of perspectives. War brings pain, loss, death, ruin, and rust. But it also provides a means for some eternal worth. And perhaps this poem, though written before his son Carol committed suicide in 1940, offers some clarity to Frost's comments on that suicide: "Two things are for sure: he was driven distracted by life and he was perfectly brave. I wish he could have been a soldier and died fighting Germany."[16]

War poetry is an expansive and not so rigidly-defined sub-genre. No one would or should try to compare Owen's "Dulce et Decorum Est" or "Arms and the Boy" or Brian Turner's "At Lowe's Home Improvement Center" or "The Hurt Locker" with any of the poems I've mentioned here. Owen's and Turner's poems fall into a separate category. But if we wonder about the conflicts we've endured and are enduring, as well as the human costs of and underlying reasons for those conflicts—the "something there is that" sends both ground swells under walls each spring and injured fathers back into battle—Frost's poems offer perspectives worth considering.

James M. Dubinsky is an associate professor of Rhetoric and Writing in the Department of English at Virginia Tech (VT). He was the founding director of both the department's Professional and Technical Writing program and VT-Engage, the University's center for civic engagement. He is a past president and the current executive director of the Association for Business Communication (ABC). Jim retired from twenty-eight years of service in the U.S. Army, and he has been instrumental in creating an academic field in Veterans Studies. Jim has been honored with awards at the college and university levels for teaching, scholarship, and outreach.

Notes

[1] James M. Dubinsky. "War and Rumors of War in Frost," *Robert Frost Journal*, no. 5, (1995): 1–22.

[2] Robert Frost and Louis Untermeyer, in *The Letters of Robert Frost to Louis Untermeyer*, 1st ed. (New York: Holt, Rinehart and Winston, 1963), pp. 285-373.

[3] Robert Frost, "Trial by Existence," in *Complete Poems of Robert Frost* (New York: Holt, 1949), 28, lines 1-4.

[4] Robert Frost, "Mending Wall," in *Complete Poems of Robert Frost* (New York: Holt, 1949), 47, lines 1-3.

[5] Frost, "Mending Wall," line 24.

[6] Frost, 48, lines 40-41.

[7] Frost, 48, lines 32-34, 45.

[8] William Wordsworth, "The Excursion," The Project Gutenberg EBook of The Poetical Works of William Wordsworth, Volume 5 (of 8), (The Project Gutenberg), accessed January 21, 2022, https://www.gutenberg.org/files/56361/56361-h/56361-h.htm, lines 672-678.

[9] Robert Frost, "The Black Cottage," in *Complete Poems of Robert Frost*, (New York: Holt, 1949), 74-77, line 31.

[10] Robert Frost, "The Black Cottage," lines 112, 117-118.

[11] Robert Frost, "Not to Keep" in *Complete Poems of Robert Frost* (New York: Holt, 1949), 284, lines 18-22.

[12] Robert Frost, "A Soldier," in *Complete Poems of Robert Frost Frost* (New York: Holt, 1949), lines 1-2. 11-14.

[13] Robert Frost, "A Soldier," line 2.

[14] Frost, "A Soldier," line 11.

[15] Frost, "A Soldier," lines 12-13, 13-14.

[16] Lawrance Roger Thompson and R. H. Winnick, *Robert Frost: The Later Years, 1938-1963* (New York: Holt, Rinehart and Winston, 1976), 73.

War Without Allegory:
World War I, Tolkien, and The Lord of the Rings

RACHEL KAMBURY

The Lord of the Rings is a war story.

It borrows heavily from that grandest of traditions set forth by works like *Beowulf* and *The Wanderer*—Old English warrior poetry, by turns heartbreaking and bloody, meant to be spoken—and its author was steeped in the same fetid waters that brewed some of the most famous novels to come out of the Great War in which he fought: *All Quiet on the Western Front, Parade's End, A Farewell to Arms* . . .

So why, in all the years since its publication, has Professor J.R.R. Tolkien's epic never found its way onto any "Best-Of" lists of war literature? Why, in light of its resonance among those who have experienced or study warfare, has it never been truly counted among the greatest novels to emerge from the events of World War I?

Because if one were to only look past the Elves, Dwarves, Hobbits, kings, and Istari, past the natural and unnatural magic, past the One Ring and its wicked maker, it would be inescapably clear that war and all that it entails rests firmly at the heart of *The Lord of the Rings.*

How could it not?

The roots of *The Lord of the Rings* broke ground during the war: Second Lieutenant John Ronald Reuel Tolkien, Battalion Signaling Officer to the 11th Lancashire Fusiliers, often found himself desperately seeking relief—however temporary—from the chaos, boredom, and brutality of war with pen strokes and scraps of paper, writing "in grimy canteens, at lectures in cold fogs, in huts full of blasphemy and smut, or by candle light in bell-tents, even some down in dugouts under shell fire."[1] It was in between his duties as officer that Tolkien began to lay the narrative foundations of what would eventually become Middle-earth. They were "fairy-stories," so-called: vignettes concerning gnomes and sprites and elf-like creatures, the kinds of stories with which Tolkien had been loosely enamored since he learned to read. It wasn't until he was invalided back to England with trench fever during the Battle of the Somme in 1916 that "Tolkien wrote out… the haunting epic of Gondolin, a city of high culture which is destroyed in a hammerblow by a nightmarish army."[2]

More stories followed shortly after; snippets of a larger world beyond the scope of any fairy-story Tolkien would have encountered in his childhood, but which would eventually become the grand mythos of Middle-earth, itself. (That mythos would not be called the *The Silmarillion* until the book's publication in 1977, four years after Tolkien's death, but its disparate pieces were brought to life both in the midst and in the aftermath of the Great War: Morgoth and the fall of Gondolin; Eä and the Undying Lands; Ainulindalë and the War of Wrath.)

A tangible causality exists between Tolkien's experiences in combat and the literature he would write and rewrite over the ensuing years of his life. War was a catalyst for Tolkien; an experience that galvanized him to write stories unlike anything he or the world around him had ever seen before. "A real taste for fairy-stories," Tolkien once explained, "was wakened by philology on the threshold of manhood, and quickened to full life by war."[3]

He wrote in another letter:

> This business began so far back that it might be said to have begun at birth . . . But the mythology (and associated languages) first began to take shape during the 1914-18 war . . . The kernel of the mythology, the matter of Lúthien Tinúviel and Beren, arose from a small woodland glade filled with 'hemlocks' (or other white umbellifers) near Roos on the Holderness peninsula—to which I occasionally went when free from regimental duties while in the Humber Garrison in 1918.[4]

Places like the Somme, once a rotten quagmire of a battlefield, now resemble any number of humble pastures, lush and green. But to walk its preserved trenches—or any others that remain in France today—is to tread a liminal space between past and present, where one can hear the souls of the dead cry out if one listens closely enough, long enough; where the scars in the earth are as deep now as they were when they were first dug and blasted; where, in a single day, more men died or were wounded than any other in the history of the British Army. He writes:

> On either side and in front wide fens and mires now lay, stretching away southward and eastward into the dim half-light. Mists curled and smoked from dark and noisome pools. The reek of them hung stifling in the still air. [Sam] fell and came heavily on his hands, which sank deep into sticky ooze, so that his face was brought close to the surface of the dark mere. There was a faint hiss, and a noisome smell went up, the lights flickered and danced and swirled. For a moment the water below him looked like some window, glazed with grimy glass, through which he was peering. Wrenching his hands out of the bog, he sprang back with a cry. 'There are dead things, dead faces in the water,' he said with horror. 'Dead faces!'"[5]

The Dead Marshes are impossible not to recognize by anyone at all familiar with trench warfare, the Somme or otherwise; pits and pools and the ghastly, decayed limbs and faces of the dead appearing as if born from muck is a hallmark of their collective memory, one Tolkien and his comrades-in-arms knew intimately. The Somme was Tolkien's first and last experience of combat, but the beginning of a years-long global event that would take the lives of almost all of his closest friends and return him—and so many others of his generation—to their own shores forever changed.

But for all of the parallels between Tolkien's lived experiences and the details in his works, he always maintained one thing: Middle-earth—its stories, its mythology, its peoples—is not an allegory. Not for his war, or any other.

In a letter to his publisher written in the waning days of 1938, while Hitler in his seat of unimpeachable power stoked anti-Semitic furor and the fires of future war, Tolkien wrote: "The darkness of the present days has had some effect on [The Hobbit]. Though it is not an 'allegory.'"[6]

Later, in 1944, days before the Allied invasion of Normandy on June 6, he wrote: "'Romance' has grown out of 'allegory,' and its wars are still derived from the 'inner war' of allegory in which good is on one side and various modes of badness on the other. In real (exterior) life men are on both sides."[7]

And again, in a letter to one Mr. Straight in 1956, he pressed again the non-allegorical nature of his works:

> I hope that you have enjoyed *The Lord of the Rings*? Enjoyed is the key-word. For it was written to amuse (in the highest sense): to be readable. There is

no 'allegory', moral, political, or contemporary in the work at all...I think that [the] fairy story has its own mode of reflecting 'truth', different from allegory, or (sustained) satire, or 'realism', and in some ways more powerful. I did not foresee that before the tale was published we should enter a dark age in which the technique of torture and disruption of personality would rival that of Mordor and the Ring and present us with the practical problem of honest men of good will broken down into apostates and traitors.[8]

Taking Tolkien's adamant refusal of all things allegorical and *The Lord of the Rings* together pushes our understanding of the novel (and it is a single novel—don't let the separate volumes fool you) into an interesting (and arguably even more valuable) place—one that rejects simplification and generalization in favor of something far more nuanced.

Tolkien wrote *The Lord of the Rings* for many reasons, chief among them to build a house for his own invented languages, and to provide Britain with a mythology besides—or perhaps beyond—the well-worn lore of King Arthur and his Knights of the Round Table. It was written with a mind toward realism—as real as elves and orcs and Nazgûl and hobbits can be—and the sense that magic is less an external force that defies nature but is in fact of nature, itself. Trees can communicate through their roots; flowers can bloom even in the most scorched earth; the sea can consume entire cities and create new lands; love can overturn the will of evil.

So if not an allegory for anything, how might we classify *The Lord of the Rings*?

Why not a war story?

Consider the Nazgûl as depicted in Peter Jackson's films, mounted on Fell Beasts and scouring the Dead Marshes for the Ring: wheeling and arching over a wretched, pocked landscape, determined to seek out and destroy its intended target, unleashing that unholy shriek which makes the blood run cold and bringing to mind the artillery shells that blasted those dank holes in the first place, at the Somme and across Europe during the Great War. An enemy that is heard before it is seen, that makes such an inhuman sound it immobilizes men, makes them panic and collapse in fear.

How can it not be a war story?

Consider Frodo Baggins, returning home to a Shire he doesn't recognize: a bucolic, green corner of the world rendered grey and flat and grim by industry and greed and hardship; how can anyone not recognize a young man returning to his home after a devastating world war, one that saw the almost immediate overthrow of agrarian life for industrialization and the loss of an entire generation? To say nothing of the feelings Frodo shares with his closest friends and confidants, Samwise Gamgee and Gandalf the Grey, at the end of his journey, which are as blatant as anything expressed in a veteran-written memoir:

> "Alas! There are some wounds that cannot be wholly cured," said Gandalf.
> "I fear it may be so with mine," said Frodo. "There is no real going back. Though I may come to the Shire, it will not seem the same; for I shall not be the same."[9]

> "Where are you going, Master?" cried Sam, though at last he understood what was happening.
> "To the Havens, Sam," said Frodo.
> "And I can't come."
> "No, Sam. Not yet anyway, not further than the Havens . . ."
> "But . . . I thought you were going to enjoy the Shire too, for years and years, after all you have done."
> "So I thought too, once. But I have been too deeply hurt, Sam. I tried to save the Shire, and it has been saved, but not for me. It must often be so, Sam, when things are in danger: someone has to give them up, lose them, so that others may keep them."[10]

Any war writing worth its ink concerns itself with war's aftermath and its repercussions, some of which can (and too often do) echo down through years and across generations, the heavy stone of the event dropped into the great sea of history and releasing ripples of trauma, instability, discord, grief.

As Tolkien himself admitted: "An author cannot of course remain wholly unaffected by his experience [...]. One has indeed personally to come under the shadow of war to feel fully its oppression."[11]

The Lord of the Rings is not an allegory for World War I. But it doesn't have to be to be of that war—born from it and in spite of it, a pupil of the same lessons Tolkien learned on the battlefield and off, through the aftermath of his own overwhelming loss. One needn't strip away the fantasy elements to make it a war novel, either; the presence of magic does not diminish its right to sit on the shelf beside Remarque and Ford and Hemingway.

The Lord of the Rings is a complex treatise on war and trauma, nature and industry, fellowship and love; the same themes shared by all great and lasting works of war literature. Poplar or willow, Nazgûl or artillery shell, marsh or trench, *The Lord of the Rings* occupies a unique place in the genre: cross-roads, if you will, where Tolkien's war and Middle-earth meet.

Rachel Kambury is a writer, novelist, and editor specializing in war and military literature and history. Born and raised in Oregon, Rachel self-published her first World War II novel, *Gravel*, in 2009—two months before she graduated high school. She earned a BA in literature from Eugene Lang College in 2013 and studied war history at the American University of Paris. Rachel's work has appeared in the *Wrath-Bearing Tree, Consequence Magazine*, the *Quivering Pen*, the United States World War I Centennial Commission, and the *Columbia Journal*. Her essay, "My First War Novel," is featured in the Military Writers Guild anthology *Why We Write: Craft Essays on Writing War*. She lives and works in New York City and can usually be found having very loud feelings on Twitter @rkambury.

Notes

[1] J.R.R. Tolkien, Christopher Tolkien, and Humphrey Carpenter, "From a Letter to Christopher Tolkien, 6 May 1944," in *The Letters of J.R.R. Tolkien: A Selection* (Boston, MA: Harper Collins Publishers, 2000), p. 78.

[2] John Garth, "Tolkien Fantasy Was Born in the Trenches," John Garth, accessed October 10, 2020, http://www.johngarth.co.uk/php/tolkien_somme_standard.php.

[3] J.R.R. Tolkien, *On Fairy Stories* (London, UK: HarperCollins, 2014).

[4] J.R.R. Tolkien, Christopher Tolkien, and Humphrey Carpenter, "To the Houghton Mifflin Co." in *The Letters of J.R.R. Tolkien: A Selection* (Boston, MA: Harper Collins Publishers, 2000), pp. 221-222.

[5] J.R.R. Tolkien, *The Lord of the Rings* (New York, NY: Houghton Mifflin Harcourt, 2013), 625-626.

[6] J.R.R. Tolkien, Christopher Tolkien, and Humphrey Carpenter, "To Stanley Unwin, 13 October 1938" in *The Letters of J.R.R. Tolkien: A Selection* (Boston, MA: Harper Collins Publishers, 2000), p. 41.

[7] J.R.R. Tolkien, Christopher Tolkien, and Humphrey Carpenter, "To Christopher Tolkien (airgraph), 25 May 1944," in *The Letters of J.R.R. Tolkien: A Selection* (Boston, MA: Harper Collins Publishers, 2000), p. 82.

[8] J.R.R. Tolkien, Christopher Tolkien, and Humphrey Carpenter, "To Michael Straight (drafts)" in *The Letters of J.R.R. Tolkien: A Selection* (Boston, MA: Harper Collins Publishers, 2000), pp. 232-234.

[9] J.R.R. Tolkien, *The Lord of the Rings*, 989.

[10] J.R.R. Tolkien, 1029.

[11] J.R.R. Tolkien, "Foreword to the Second Edition" in *The Lord of the Rings* (New York, NY: Houghton Mifflin Harcourt, 2013), xxiv.

The October Revolution, Russia Occupation of Persia: WWI Soldier Viktor Shklovsky's A Sentimental Journey: Memoirs, 1917-1922

MICHAEL CARSON, veteran

After the explosion our soldiers, surrounded by enemies, were waiting for a train to come for them; while waiting, they busied themselves by picking and putting together the
shattered pieces of their comrades' bodies.
They picked up pieces for a very long time.
Naturally, some of the pieces got mixed up.
One officer went up to a long row of corpses.
The last body had been put together out of the leftover pieces.

It had the torso of a large man. Someone had added a small head; on the chest were small arms of different sizes, both left.

The officer looked for a rather long time; then he sat on the ground and burst out laughing ... laughing ... laughing ...

–From Viktor Shklovsky's *A Sentimental Journey: Memoirs, 1917-1922*[1]

Why read Viktor's Shklovsky's *Sentimental Journey: Memoirs, 1917-1922* a hundred years after the First World War? Why remember this account of the October Revolution and the Russian occupation of Persia when we have forgotten so many other accounts of the First World War in favor of those charnel-house memories of gallant-British officers at the Somme and Ypres? What does this young Russian commissar have for us today except for yet another account of yet another endless bloody war that few remember now, and no one will remember in a hundred years?

For one, *A Sentimental Journey's* perspective on WWI is unique and difficult, not simply because of its various, diverse, and, from an Anglo-American perspective, obscure experiences but also because of its form. Shklovsky writes in stilted sentences, delays information, mixes up chronology. He claims he only wants to report the facts. He wants to become a primary source. But he insists the facts must be reshuffled, drawn out, ironically juxtaposed, removed from their logical spot in one paragraph and placed at the end of the next. He has a terrible memory. Here is Shklovsky on his brother's death:

> He cried hard before dying.
> Either the Whites or the Reds killed him.
> I don't remember which—I really don't remember. But the death was unjust.[2]

A leitmotif, ignorance and naivety—a sort-of forced dramatic irony—winds through *Sentimental Journey*, disturbing and disrupting moments of solemnity and despair, moments that, as Cassandra says before her fated death in Aeschylus' *Agamemnon*, that truly break the heart.[3] Here is Shklovsky in Persia, a commissar in the Army, tasked with pacifying the local population and keeping the Russian forces there from mutiny:

> One morning when I got up and opened the street door, something soft fell to the side. I stooped down and looked . . . Someone had left a dead baby at my door.
> I think it was a complaint.[4]

Later Shklovsky demands through telegram that the Russian government withdrawal the Russian units stationed in a foreign country:

> Have inspected Kurdistan units. In the name of revolution and humanity, demand withdrawal of troops.
> This telegram didn't go over too well—apparently, it's naïve and funny to demand the withdrawal of troops in the name of humanity.[5]

Occasionally the chain of ignorance gives way to clarity, as when he explains why the Petersburg garrison played such a prominent role in the February Revolution, contradicting pretty much every historian's account of the insurrection:

> This may be childish, but I'm convinced that restriction to the barracks, where men torn from their duties rotted on bunks with nothing to do, the

dreariness of the barracks, the dull despair and resentment of the soldiers being hunted down on the street—all this stirred up the Petersburg garrison more than then constant reversals in the war and the persistent rumors of "treason."[6]

His summary of the Russian experience in Persia will resonate with anyone who has taken part in ill-advised imperial projects. He says of his time in Persia:

We had gone to a foreign country, occupied it, added to its gloom and violence our violence, laughed at its laws, hampered trade, refused to let it open any factories and supported the shah. And for this purpose we kept troops—kept them there even after the revolution. It was imperialism—what's more, Russian imperialism, which is to say, stupid imperialism.[7]

And then the question at the heart of this *Journey:* what good is memory without responsibility? He writes:

All our shrewd and far-sighted policies were for nothing. If, instead of trying to make history, we had simply tried to consider ourselves responsible for the separate events that make up history, then perhaps this wouldn't have turned out so ludicrously.[8]

Like Robert Graves' *Goodbye to All That, A Sentimental Journey* is a sad book with much hysterical laughter. Graves dismissed his own memoir as a cobbled together hack job:

People like reading about food and drink, so I searched my memory for the meals that have had significance in life and I put them down . . . People like hearing about T.E. Lawrence, because he is supposed to be a mystery man . . . people like reading about poets. I put in a lot of poets.[9]

For Shklovsky the cobbled together nature of the whole is not like Graves' marketing technique, a satire on the system that made it possible. The cobbling together is no by-product or accident; it is the most important point of *A Sentimental Journey*'s structure.

Shklovsky was a formalist. He believed art had certain forms, patterns. The artist rearranged these patterns, delayed them, inverted them. "Formalism," he said, "does not deny the idea content of art, but treats the so-called content as one of the manifestations of form."[10] His sentimental journey (itself a re-patterning of Lawrence Stern's novel of the same name) takes the lopped off limbs of people and history and reconstitutes bodies. It places them in the wrong place. It makes the reader see the whole by making it un-whole, by distorting the corpse, disturbing to revivify, to, as he explains in his *Theory of Prose,* "make the stone feel stony." He called this process "ostraniene" or "estrangement." All writers worth reading estrange in one way or another, "return sensation to our limbs" by making what was familiar unfamiliar and alive.[11] If they do not, they give the reader what they already have; they deliver them into the same old plots—the same old histories, the same old wars. They add more dust.

This writing process did not make Shklovsky many friends, either in the world of artists or those of memoirists. It has made him even fewer friends among historians and the later literary theorists who abandon language entirely for structure. But it has the advantage of being true. Artists—whether writing about war or not—do succeed in so far as they can manipulate forms of language, and their success depends less on the force of their content but how that content is manifested in form, in how assiduously they study and internalize the movements of syntax, the historiography of plot. Formalists resist, as Gerald L. Burns argues in an essay on Shklovsky, "the urge to experience an object as something other than it is;" they make the "stone feel stony" and consequently realize the truly subversive truth and possibility of art—that it does not merely imitate, that it can at its best, as Shklovsky contends in his *Theory of Prose,* "incite insurrections among things."[12]

But the content cannot speak for itself. Shklovsky dismissed most accounts of the war. He mocked Barbusse's 1916 novel of the trenches. "Contrived," he said. "Glossy." Barbusse's novel did not have a sense of what happened—"with its jumble of corpses, its end washed away by various conclusions."[13]

Shklovsky understood that writing about war, and writing generally, must not be a jumble of corpses; life might be random, but art must be, above all, precise, meticulous. This is what the artist does. This is what a human can do in the face of the immensity and horror of so much relentless blood and history, so many disproportionate parts, so many competing claims to truth: they cannot give themselves over to representations and interpretations that dismiss through explication; they must both respect the thing itself and the artists role in manipulating the thing; they must make the stone feel stony." He says toward the end of *A Sentimental Journey:*

When you cut off a leg, You have to cut through the muscles, pull back the flesh with forceps and saw through the bone.

Otherwise, the bone will eventually pierce through the stump.

If you don't like this description, don't make war. As for me, I'm ashamed to walk down the streets of Berlin and see the cripples.[14]

Shklovsky suggests that we might have little control over the sweep of history, but we do have a responsibility to represent the particular and then to rearrange the particulars for maximum effect; memory in and of itself is nothing—it is all about how the writer arranges the memory; how she organizes the pieces; how the audience responds to the rearrangement.

But still. Why this particular account of war? Why should we remember the Russians in Persia when there were bigger battles? Bloodier massacres? Because "nowhere," Shklovsky tells us, "was the inner lining of war, its predatory essence, so clear as in the crevices of Persia," and it was clear, he argues, because "there was no enemy";[15] only there, in the sad corners, where forgotten armies disintegrated, starved, and marauded, where ancient enmities yielded slaughter and counter-slaughter, could you see the war, see war, for what it was.

And Shklovsky's war—despite the memoir's title—is not a sentimental one. His brother is killed, and he does not remember who killed him. A bomb explodes in Shklovsky's hands, and he gives next to no detail as to the consequences. There are no enemies here, no friends. "No," Shklovsky says after describing the American ambassador to Persia, Doctor Shedd, trying to get as many Aissor children inside his surrey, to protect the youth from the general butchery, to deliver them back to their fathers, "I shouldn't have written that. It warmed my heart. It . . . aches."[16]

Unlike the American ambassador, Doctor Shedd, trying to get those children into the surrey, Shklovsky did not have a bucolic home and an orderly small town to forget the far-flung horrors of empire. He did not have the luxury of sentimentality. He was caught in the horror, both perpetrator and victim, both criminal and innocent ("we don't count corpses in the East," he tells the ambassador).[17] Shklovsky was like America a hundred years later, like America in Vietnam, Iraq, Afghanistan, and Nigeria where the innocence of Doctor Shedd is only a memory, a pleasant memory, but a memory nonetheless.

But Shklovsky did have art. He did have the skill to make those corpses and human life something other than numbers, something other than a lie. He could make something comic and horrible and lovely out of leftover body parts. He could hold a bomb in his hands and watch it explode.

Vladimir Nabokov, another survivor of this wretched period of Russian history, caught like Shklovsky in a war that had no front, no rear, no perspectival dichotomy to control and assuage the violence like their Anglo counterparts, also like Shklovsky exiled, once wrote a novel that accomplishes exactly what Shklovsky said art should accomplish; Lolita's inspiration, Nabokov claimed, was an ape in the Jardin des Plantes, who, after months of coaxing by a scientists, "produced the first drawing ever charcoaled by an animal: this sketch showed the bars of the poor creature's cage."[18]

One more quote from Shklovsky:

Of course, I'm not sorry that I kissed and ate and saw the sun. I'm sorry that no matter how hard I tried to direct events, they went their own way. I'm sorry that I fought in Galicia, that I got mixed up with armored cars in Petersburg, that I fought along the Dnieper. I changed nothing. And now, as I sit by the window and look at the spring, which goes past me without asking what weather it should arrange for tomorrow—which doesn't need my permission, perhaps because I'm not from around here—I'm thinking I should have probably let the revolution go past me in the same way.

When you fall like a stone, you don't need to think; when you think, you don't need to fall. I confused two occupations.

The forces moving were external to me.

The forces moving others were external to them.

I am only a falling stone.

A stone that falls and can, at the same time, light a lantern to observe its own course.[19]

Shklovsky steals this from Baruch Spinoza. But what is art but the manipulating of old forms? Old ideas? Ancient juxtapositions? Enduring analogies? What is art but the moving around of the limbs we have left to us? What do artists do but—as Shklovsky contends in his Energy of Delusion—"convert both torment and joy into experience?"[20] Transform and transmute this torrent of parts into content? Into life?

Few talk of Shklovsky anymore. Few talk of the Russian Persian garrison of 1917. Few know what to do with the leftover pieces. It's just the way of things.

There are new writers. New wars. New bodies. We continue to fall, trying, with poor Prufrock, to hear voices dying "beneath music from another room."[21] But we can still hold up the lamp, we can manipulate the forms, we can see our plummeting stone, illuminate the bars of our cage, convert the horrible sweep of long ago and ongoing violence into experience—clean the dead world, make it alive, make it hum, turn it to art.

Michael Carson served for four years in the United States Army Infantry, including a 13-month 2006-2007 tour of duty in Mosul, Iraq. He now teaches in Baytown, Texas, and helps edit the *Wrath-Bearing Tree,* a nonprofit literary journal and imprint. His non-fiction has appeared at *Numéro Cinq* and the *War Horse*, and his fiction in the short story anthology *The Road Ahead: Stories of the Forever War.*

Notes

[1] Viktor Shklovsky, *A Sentimental Journey: Memoirs, 1917-1922*, trans. by Richard Sheldon (Champaigne, Ilinois: Dalkey Archive, 2004), 126.

[2] Shklovsky, *A Sentimental Journey,* 156.

[3] Aeschylus, *The Orestia*, trans. by Richard Fagles (New York: Penguin Books, 1979, 158.

[4] Aeschylus, *The Orestia*, 115.

[5] Aeschylus, 101.

[6] Aeschylus, 7.

[7] Aeschylus, 78.

[8] Aeschylus, 102.

[9] Robert Graves, *Robert Graves* (New York, 1961), 68-70.

[10] Shklovsky, *A Sentimental Journey,* 232.

[11] Viktor Shklovsky, *Theory of Prose* (Champaigne, Illinois: Dalkey Archive Press, 1991), 6.

[12] Gerald R. Burns, Introduction to *Theory of Prose* (Champaigne, Illinois: Dalkey Archive Press, 1991), xi.

[13] Shklovsky, *A Sentimental Journey,* 156., 59.

[14] Shklovsky, 220.

[15] Shklovsky, 78.

[16] Shklovsky, 276.

[17] Shklovsky, 270.

[18] Vladimir Nabokov, "On a Book Entitled Lolita," *Encounter* (April 1959), 73-76.

[19] Shklovsky, *A Sentimental Journey*, 133.

[20] Viktor Shklovsky, *Energy of Delusion: A Book on Plot*, trans. by Shushan Avagyan (Champainge, IL: Dalkey Archive Press, 2007, 150.

[21] T.S. Eliot, "The Love Song of J. Alfred Prufrock," *Collected Poems: 1910-1962* (New York: Harcourt Brace & Company, 1991), 3.

*Army of Shadows**

ROXANA ROBINSON

I read only one war novel while I was writing my own. There were reasons: I didn't want to hear another novelist's voice, as I was trying to find my own way into a soldier's mind. I didn't want to learn about the wrong war; every war has its own weather and terrain, its own equipment and language, and I didn't want the wrong ones in my head. I couldn't read novels about the Iraq war because there were none.

War writing follows a sequence: The first writers are reporters. Later, veterans publish memoirs. Novelists are always last; ten years after the invasion, the first novels about the Iraq war appeared. Fiction is ruminative, emerging slowly from experience, like water seeping upward to a spring.

My own book, *Sparta*, was engendered by an account in the *New York Times* about unarmored Humvees, about IEDs and TBIs. It told of the reluctance of the military medical establishment to diagnose brain injuries, because their treatment was expensive, and would remove combatants from the field.

The clarity of the writing, the intimacy of the images, and the urgency of the issues collapsed the distance I'd kept between myself and the war. I had opposed it, but it had happened. I felt helpless and didn't want to think about it.

This article made that impossible. Those soldiers stayed in my mind, taking more and more space, until I realized that this would become a book.

I knew almost nothing about war—I'm a Quaker. To write a book about this I needed a universe of information. I read reportage and memoirs. On the internet, I read military blogs by soldiers in Ramadi and Hit and Fallujah. On You-Tube I watched Marine dance contests and family homecomings. I watched real firefights: soldiers put babycams on their helmets and switched them on when the shooting started. I interviewed every vet I could find. I listened to their stories. And I asked them their favorite war books.

The problem of war fiction is scale. War is vast, abstract and impersonal. It may be the most impersonal of all human endeavors; if we had to consider it in personal terms we would never wage it. A soldier is small, human and personal. The task is to create a narrative that includes the personal and impersonal, and to create, in a language that civilians will understand, a world they can't imagine.

I kept hearing one title.

All Quiet on the Western Front, by Erich Maria Remarque, was published in 1929, about a decade after WWI. Drawing on the author's experience, it's a first-person narrative, told by a young German soldier fighting in France.

Remarque beautifully resolves the problems of scale and language. Private Paul Baumer's voice is quiet and informal, and the story begins with a matter of universal interest: food. "Yesterday we were relieved, and now our bellies are full of beef and haricot beans."[1]

Not only the soldiers are relieved; the readers are as well. Braced for horror, we've gotten a reprieve. We needn't face the big guns just yet; we're safe behind the lines, and replete. Soon we learn why this has happened; most of the company has been killed. The cook prepared food for one hundred and fifty soldiers, but only eighty have returned from the front lines. This news is delivered casually; it's important mainly because the surviving soldiers want all the food.

The sentimental notions of war and heroism are challenged at once, and the narrative continues to shock us with its contrasts between dailiness and horror. But the voice is not callous; in fact, it's almost unbearably perceptive and compassionate. Bauer visits poor Kemmerich, dying of blood poisoning in the field hospital. Kemmerich is told he's going home, and he nods without speaking. Bauer says, "I cannot bear to look at his hands, they are like wax. Under the nails is the dirt of the trenches, it shows through blue-black like poison."[2]

The life of the body, in all its exigence, rises up through the narrative like a skeleton through earth. As death approaches, life becomes more urgent; men wolf down their food before going out, because they might die before another meal. A wounded comrade dies in No Man's Land because he's blinded and mad with pain, blundering into the gunfire before his friends can reach him. It's the intimate moments that inform us of the ghastly presence of war, looming behind everything. It's war that makes poor Kemmerich's hands so waxy, war that causes the clumsy rush into machine-gun fire.

The narrative is both beautiful and desolate. It's shocking, in the way of great fiction, because it reveals things we knew but didn't know we knew. It's modern in its rejection of conventional pieties about courage or nobility or patriotism. It's complete in its grasp of the wartime experience, with all its futility and heartbreak, extremity and complexity, its moments of deep human connection.

Two things serve as testaments to the book's merit: that American veterans so admire it, nearly a century later, and that the Nazis were so enraged by it, and its rejection of their notion of military nobility, that they burned it and banned it, and then, since Remarque himself was unavailable, they decapitated his sister.

Books have a kind of power that war cannot equal, but war has a kind of power that seems limitless and unending. It was that paradox I wanted to explore; it was that paradox that all war books, by their nature, challenge and perpetuate.

*This essay was first published in *Bookforum* in June/July/August 2014. Reprinted with permission of *Bookforum*.

Roxana Robinson is the author of ten books—six novels, three collections of short stories, and the biography of Georgia O'Keeffe. Four of these were chosen as *New York Times* Notable Books , two as *New York Times* Editors' Choices. Her fiction has appeared in the *New Yorker,* the *Atlantic, Harper's, Best American Short Stories, Tin House,* and elsewhere. Her work has been widely anthologized and broadcast on NPR. Her books have been published in England, France, Germany, Holland, and Spain. She is the recipient of many awards, the most recent the Barnes & Noble "Writers for Writers" Award, from Poets & Writers.

Notes

1 Erich Maria Remarque, *All Quiet on the Western Front,* trans. A.W. Wheen Fawcett Crest, https://www.pdf-archive.com/2016/05/12/aqwf-full-text/aqwf-full-text.pdf, pdf, p. 2.

2 Erich Maria Remarque, *All Quiet on the Western Front,* 8.

Lucien and Albert Camus: WWI Father, WWII Resistance Son

JENNIFER ORTH-VEILLON, PhD

Albert Camus' mother, Catherine Hélène Sintes, occupies a central role in the French writer's life and work. Much less had been known about his father, Lucien Auguste Camus until Camus' untimely death in 1960 at forty-six years old. Found among the wreckage of the car accident that killed him, was his unfinished novel, *The First Man,* which tells the story of a man's pilgrimage to discover himself through family and cultural history. While fictional, the book draws upon Camus' autobiography—his poor Algerian childhood, his eternal feeling of estrangement from France, and a quest to know more about the father he never met, Lucien. Lucien Camus, who had spent his entire life in Algeria, was injured at the Battle of the Marne in the early months of WWI and died in France on October 11, 2014 at a hospital in St. Brieuc.

In his notebooks, Camus perhaps mused about writing *The First Man* when he declared, "The work is a confession. I need to testify."[1] Through the imagination of the narrator-character of the son, Jacques Cormery, *The First Man* bears witness to the terrible butchery of WWI that takes the life of the father character, Henri Cormery. The second chapter begins as Jacques travels to his father's grave near Saint-Brieuc in Brittany. Jacques recalls that the Saint-Brieuc hospital sent his mother, Lucie Cormery, the piece of shell that wounded his head along with his father's last words. Camus writes:

> The piece of shell that had torn open his father's head was kept in a small cookie jar behind the same towels from the same dresser, with the cards written from the front whose words he could recite by heart in their

dryness and brevity. "My dear Lucie. I am well. We are changing quarters tomorrow. Take care of the children. I love you. Your husband."[2]

Another note read, "I am injured. It's nothing. Your husband." He died a few days later. The nurse had added a note to the death notice, which read, "It's better this way. He would have stayed blind or crazy. He was very courageous."[3]

When M. Bernard, Jacques' beloved primary school teacher, paternal figure, and a veteran of WWI, reads to his class aloud from Roland Dorgelès' WWI masterpiece, *Wooden Crosses*, it's as if Jacques bears witness to his father's combat experience at the Marne:

> ... the Zouaves, he said, they were put out in front and then at the charge down a ravine they charged and there was no one ahead of them and they were advancing and all of a sudden the machine gunners when they were halfway down they were dropping one on top of the other and the bed of the ravine was all full of blood and the ones crying for *maman* it was awful.[4]

M. Bernard reads until he reaches the story's end where the character, D. gets killed. The teacher closes the book and raises his eyes to see Jacques "in the first row staring at him with his face bathed in tears and shaking with sobs that seemed as if they would never end."[5]

However, when the forty-year-old Jacques stands in front of his father's grave at St. Brieuc decades later, he feels no emotion. It's only when he realizes he is eleven years older than his father was at the time of his death, he becomes

> ... struck by an idea that shook his very being ... And the wave of tenderness and pity that at once filled his heart was not the stirring of the soul that leads the son to the memory of the vanished father, but the overwhelming compassion that a grown man feels for an unjustly murdered child– something here was not in the natural order but, in truth, there was no order but only madness and chaos when the son is older than the father.

The sense of injustice makes Jacques dizzy, and he admits that "the statue that every man eventually erects and that hardens in the fire of the years, into which he then creeps and finally awaits its final crumbling—the statue was rapidly cracking, it was already collapsing." Jacques looks back on his life, ashamed of being "foolish" and "cowardly," aiming toward "a goal which he knew nothing about." The father who had always been a stranger suddenly appeared real and important to Jacques: " . . . the secret he eagerly sought to learn through books and people now seemed to be intimately linked with this dead man, this younger father."[6] He's seized with a desire to know him better, but just when Jacques is about to dig further, he understands "there was nothing more for him to do here." An airplane passes overhead, breaking the sound barrier and he walks away: "Jacques Cormery abandoned his father."

Jacques Cormery may have felt he'd forsaken his father in this abstract sense, but when the Germans invaded and occupied France in 1940, Albert Camus did share something with Lucien: he fought. Camus was too weak from tuberculosis to take on the armed fight, but he did risk his life by joining the French Resistance as a writer. He edited the underground journal *Combat* alongside Jean-Paul Sartre and penned some of the most influential texts from the time. Shortly after the war, he published *The Plague*, which, besides its accurate psychological and logistical descriptions of a pandemic, remains a powerful literary representation of France during the German Occupation.

Camus' efforts gain importance when put into historical context; he joined only a handful of intellectuals in the fight. Most did nothing and some, like WWI veterans Louis-Ferdinand Céline, Maurice Pujo, and Maxime Real del Sarte, were Nazi sympathizers, Vichy supporters, and anti-Semites. Others like Marc Bloch, Louis Aragon, or René Cassin resisted precisely because they fought for France in 1914-1918. WWI veterans survived the trenches only to be assassinated or deported as resistors, and many were betrayed not by the Germans but also by their own compatriots. Marc Bloch, one of France's greatest medieval historians, recipient of the *Croix de Guerre* and the *Légion d'honneur* during WWI, and a key figure in decoding letters for the French Resistance in Lyon, was arrested and brutally assassinated by the Nazis just two months before the liberation of France.

In August 1944, Camus dreamt of a just France regenerated by a democratic revolution, which, at the time, he believed came from the solidarity and sacrifice demonstrated by the French Resistance. He wrote in *Combat*, "It is only at this price that France will resume the pure face that we have loved and defended above all else."[7] Yet that pure face never showed itself to Camus during his lifetime. He would always feel ostracized intellectually from the country he defended and ambivalent about its political policies. A decade later, France would find itself at war with Algeria, a war where Camus would find himself in the terrible, contradictory position of explaining that he supported equal rights in the colonies and abhorred the war waged against them, but he remained opposed to independence.

When Jacques Cormery in *The First Man* turned his back on his father's war grave, perhaps he wasn't turning his back on his father as he declared at the end of the chapter. It's possible he was turning his back on war itself—on its violence, destruction, and absurdity—as an impossible and inhuman way to relate to him. *The First Man* searches for lost pieces of a man's identity, but it also exposes and denounces the way that war frays the filaments of a family's life across generations.

Notes

[1] Albert Camus, *Œuvres complètes II* (Paris, France: La Pléiade, 2006), 795. My translation.

[2] Albert Camus, *The First Man*, trans. David Hapgood (New York: Vintage, Random House, 1996), chap.5, Kindle.

[3] Albert Camus, *The First Man,* chap. 5, Kindle.

[4] Camus, chap. 6A, Kindle.

[5] Camus, chap. 6A, Kindle.

[6] Camus, chap. 2, Kindle.

[7] Albert Camus, "À guerre totale résistance totale," *Combat*, mars 1944, in *Œuvres complètes* I (Paris, France: La Pléiade, 2006) pp. 911-913. My translation.

part nine

POETIC RESPONSES TO WWI

Have You Ever Tried to Write a War Poem?

FALEEHA HASSAN

Have you ever tried to write a war poem? Some people believe that writing poetry requires training and planning. That if you go to school and attend poetry-writing classes or get experience by attending workshops, then you will get the ability you need to write poetry.

Honestly,

I am not totally against this idea, but there are some questions that jump into my head every time I hear it.

Such as . . .

Is *everyone* capable of writing poetry even if s/he does not have natural talent?

Are training and planning alone sufficient for the birth of a poem?

Even if we are convinced that training and planning are generally necessary, writing a poem about war is completely different. Going to school or workshops to gain enough experience to write a poem about war is not enough. No matter what we do to prepare, war poems remain frightening at their birth.

Let's explore that idea.

Let's say you are sitting on your rocking chair on your balcony on a beautiful quiet morning. As usual, you are smoking your cigarette. Next to you is a glass table with a cup of coffee. Suddenly you decide to write a poem. To be specific, you tell yourself, "Today I will write a war poem."

You take a deep breath and a sip of your coffee. You close your eyes to imagine a short war scene of a little girl sitting on the floor trying to hide her head between her knees. She is shaking and does not dare turn around for fear of the sound of the fighter planes. She panics as the sound gets louder.

Or . . .

You might imagine a group of children without shoes screaming and running in every direction, looking for a safe place to shelter. They hear the warning siren and stumble into the holes in the streets made by tanks.

Now, what do you think?

Is it easy for you to find the short, concise phrases that show how you feel and what you sense to describe the panicked little girl?

Is it easy for you to find a convincing, persuasive phrase to help the reader visualize the status of those frightened children?

I honestly do not think you could write those phrases unless you were that little girl who was hiding during the raid. That girl who was experiencing with all her senses the death that was approaching her moment by moment. Or you would have to be one of those children who was running in every direction, hoping to survive?

Only then could you write a real war poem.

And in that moment—the moment of writing the war poem–you will remember how you miraculously survived the shrapnel and the missiles. How the sound of the guns and engines of the fighter craft would not go away from your ears. And you will find out that in that moment that you are reliving those dangerous memories, everything around you will turn into a war zone.

Suddenly the chirp of a bird turns into a warning siren. You see the dove that is flying over your head become a military observation plane. The tangled bushes on your balcony become barbed wire. The air almost fades around you, and the space closes in and tightens.

You want to scream but you can't, because your voice is now mingled with the muffled murmurs of that shaking girl and those frightened children.

Even your cigarette becomes a dynamite finger that is about to explode in between your real fingers. So dangerous that you need to carefully put it in the ashtray. The taste of your coffee becomes bitter.

You try and try again to find a gap through which to escape from your memories that have now turned into horrible present moments. And when you do not find that gap, different feelings start to conflict inside you. You want to cry, but there are not enough tears in your eyes to wash your soul of that terrible moment—a moment which is full of fear.

But you try and try and try anyway to find a moment of peace like you felt this morning, not so long before your poem was born—after a very difficult labor.

Yes, writing a war poem is a dangerous process. You may lose your sense of internal peace, or you may lose that moment of joy that you feel when you notice the small things around you.

Maybe some who read this article will say that the writer is a plaintiff, or that her words are exaggerated, or that she is insane. But I say to all of them: Whether it's WWI or Iraq, no one can write a war poem where the reader can truly hear and see and feel the war moments inside it unless s/he has experienced sitting on the edge of death. And has been lucky enough to survive and become a poet.

Faleeha Hassan is a black Iraqi author born in Najaf, Iraq, in 1967, who lives in the United States. She earned an MA in Arabic literature and published several collections of poetry in Arabic: *Being a Girl, A Visit to the Museum of Shade, Five Titles for My Friend—The Sea, Though Later On, Poems to Mother, Gardenia Perfume*, and *The Guardian of Dreams*. Her works of Arabic prose include *Hazinia* or *Shortage of Joy Cells* and *Water Freckles*. Her poems have been translated into English, Italian, German, French, and Kurdish. Hassan, author of *War and Me*, has received awards from the Arab Linguists and Translators Association (WATA) and the Najafi Creative Festival for 2012, as well as the Naziq al-Malaika Prize, the al-Mu'tamar Prize for poetry, and the Shaheed al-Mihrab Foundation's short story prize.

All the Way Home*

JANE CLARKE

Against barbarity, poetry can resist only by cultivating an attachment to human fragility, like a blade of grass growing on a wall as armies march by.

–Mahmoud Darwish (1942-2008)

When the Mary Evans Picture Library in London invited me to write a sequence of poems in response to a First World War family archive, I was initially hesitant. Poems come of their own volition; I didn't know if I would find the emotional resonance needed to spark not just one poem but a series.

However, the photographs and the letters between brother and sister drew me to them, evoking poems about what it is to sacrifice a life and what it is to live with loss.

Albert Auerbach joined the British army on September 1, 1914, aged twenty, and died at the Somme four years later to the day. He and his elder sister, Lucy, were close and his many letters to her gave me insights into both their lives. A poignant line from one of these became the epigraph for the book: "It will all be over one day, and what a day it will be." As the poems accumulated, I saw a story emerging in their voices—from an evening in a September garden before Albert joined up in 1914 to Lucy's pilgrimage in 1920 to the trenches where he died.

The work I loved by Wilfred Owen, Francis Ledwidge, Siegfried Sassoon, Isaac Rosenberg echoed loudly. I had to abandon the first flush of poems because they were too clichéd or derivative, using over-statement and worn-out imagery. I searched for a more allusive approach. The first poem that worked, "After We're Gone," was inspired by an account in Richard Mabey's *Weeds* of the flower and vegetable gardens planted in the trenches.[1] In *Where Poppies Blow*,[2] John Lewis-Stempel writes about soldiers' detailed descriptions of the surrounding landscape, birds, plants and animals in their letters and poems. Their affinity with the natural world helped sustain them, humanizing an inhuman existence; this gave me a way into the sequence.

In one of the photographs Lucy's standing close to a Jersey cow on the farm in the Malverns where she helped out as part of the war effort. I wondered what it was like for a young woman brought up in a middle-class London family to learn how to milk a cow. Remembering my own attempts as a child, the thought of her struggling to find the necessary combination of strength and rhythm in her fingers led to a poem entitled "Milk." Another poem, "Axe," looks at change in women's lives because of their contribution to the war.

Lucy was a gifted pianist and went on to study with the celebrated Myra Hess. When I read that Albert was invalided home with shellshock and dysentery, I wrote a poem in which tuning and playing the piano is a metaphor for her wish to soothe his anguish. In Albert's last letter home on August 18, 1918, he thanks Lucy for sending a sprig of heather. Heather thrives in the mountains around us in Wicklow. I wondered what the heather held for her and what it meant to him to receive it; this led to a poem entitled, "Ling."

A visit to an exhibition in the National Library, Dublin in July 2017 revealed a rich archive of letters between Irish soldiers on the Front and their families at home. The Irish experience of the First World War was largely overlooked and even denied until relatively recently. Now we know that 210,000 Irish soldiers fought and up to 40,000 died. Imagining Albert and an Irish soldier lying in the same Casualty Clearing Station led me to rework an earlier poem into "Mortal Wound."

Mortal Wound

The gunner recites his village
easing in for the night –

horse chestnut trees circling the green,
children on swings,

their mother's shout, *come in
out of that or the púca will catch you,*

a couple sitting close on a bench,
oblivious to everything,

someone calls *quiet* for the weather forecast,
no stars to be seen,

but sure as a loaf on the back of the stove,
the moon slowly on the rise.

In that same exhibition I saw a photograph of people in a bog picking moss. This set another poem ticking, "Base Hospital, Boulogne." I was familiar with sphagnum moss from walking in the mountains, but I hadn't known about its powerful antiseptic qualities or that tons were gathered in Irish bogs, made into at least a million dressings, and sent to military hospitals throughout the war. Another poem was inspired by the eyewitness account of a football match on St. Stephen's Day 1916 by Fr. Frank Browne, Jesuit chaplain to the Irish Guards. He wrote, "This is just one little incident of the war, showing how little is thought of human life out here. It sounds callous, but there is no room for sentiment in warfare, and I suppose it is better so."[3]

Among the many books I read over the eighteen months working on the sequence, Vera Brittain's *Testament of Youth* stood out. [4] A scene in which she describes seeing her fiancé's kit returned after he died at the Front helped me imagine Lucy's response to her brother's death in a poem entitled "Snow."

While I was learning about the role of nationalism, imperialism, and militarism in the lead up to the First World War, nationalism was on the rise again all over Europe. Most European countries were doing all they could to avoid taking in refugees from war in Syria with a largely xenophobic public discourse about what we Europeans would lose rather than what we could gain. A visit to the *In Flanders Fields Museum* in Ypres highlighted the vast numbers of people displaced by all wars. This led to a poem about a Belgian refugee. Meanwhile Brexit was threatening the relatively fragile reconciliation in Northern Ireland and the maturing relationship between Ireland and the UK. It mattered to me that *All the Way Home,* with its twenty-one poems and eleven photographs, was coming into being through collaborative relationships across the Irish Sea, with the Mary Evans Picture Library in Greenwich, Albert and Lucy's niece, Patricia Aubrey in Ealing, and the publisher, Smith|Doorstop, in Sheffield.

Poems respond to stories, tell and retell stories, and also unearth stories we have buried. When I read from *All the Way Home,* audience members often speak about their family's involvement in the war, experiences that were unspoken in Ireland for generations. Last November, having heard a radio interview about my book, my uncle asked me if I knew that my great-grandmother had lost two brothers in the Battle of the Somme. I must have heard this as a child but somehow I had forgotten, a memory erased like so many other Irish memories of the First World War. It seems to me now that in responding to Lucy's loss I was also responding to my own great-grandmother Abby's loss and for this I am deeply grateful.

All The Way Home was published by Smith|Doorstop in April 2019. Poems reprinted with permission of the author.

Irish poet **Jane Clarke** is the author of two poetry collections, *The River* and *When the Tree Falls* (Bloodaxe Books 2015 & 2019), as well as an illustrated chapbook, *All the Way Home,* (Smith|Doorstop 2019). Jane's awards include the 2016 Hennessy Literary Award for Poetry and the 2016 Listowel Writers' Week Poem of the Year. *The River* was the first poetry collection to be shortlisted for the Royal Society of Literature's Ondaatje Prize. She holds a BA in English and Philosophy from Trinity College, Dublin, and an MPhil in Writing from the University of South Wales, and has a background in psychoanalysis. Jane lives with her wife in Glenmalure, Country Wicklow. www.janeclarkepoetry.ie

Notes

[1] Richard Mabey, *Weeds: The Story of Outlaw Plants* (London: Profile Book, 2012), 201-4.

[2] John Lewis-Stempel, *Where Poppies Blow: The British Soldier, Nature, The Great War* (London: Weidenfeld and Nicolson, 2017).

[3] Fr. Frank Browne December 1916 letter to his Jesuit provincial in Ireland, quoted in Eddie *O'Donnell, Father Browne's First World* War (Dublin: Messenger Publications, 2015)

[4] Vera Brittain, *Testament of Youth* (London: Virago Books, 2014)

The Land Remembers and Zone Rouge*

AMALIE FLYNN, PhD

THE LAND REMEMBERS

Trenches that twist and turn like empty veins. Bomb craters. Chemical foliage forests. Vapor cities. Burn pits. Dust cities. Radioactive rivers. Dioxin dug into the ground. Unexploded ordnance. Landmines strewn. Alive and circular like metal breasts. A child standing on top of a pile of what is left behind. When war floods a place. The land remembers. It holds war and keeps it. War sinks in a river. Accumulates in the gill of a poisoned fish. Laces itself in the soil and sediment. Twists inside trunks of trees. The land remembers.

Or how it bears witness to our forgetting.

And when war flooded France. The fields and the forests of Verdun. It drenched the land with bodies and bombs and blood. Or after. How the government marked the land destroyed. A no-go zone. Because war sank into that land. And never went away.

A century later.

How war is still there.

ZONE ROUGE (a poem for the centennial)

1.
When the land was.

2.
Full of bodies dead. And twisted.

3.
When the fighting was.

4.
Sustained.

5.
With bodies. Dead. Twisted on a riverbank.

6.
Wrist bent. Hand hovers. Over water.

7.
Dead bodies with fingers. Like feathers.

8.
Stretched feathers or the calamus.

9.
Attaching to bird skin.

10.
These are bodies. Bodies of war.

11.
Dead with. Feathered fingers.

12.
Wing of a bird.

13.
300 days of shelling.

14.
The shells were 240 mm. Full of shrapnel.

15.
Mustard gas.

16.
Hitting men and hitting ground.

17.
Making holes. Upon impact.

18.
Shrapnel bursting.

19.
Bloom and rip.

20.
Ripping through dirt and faces.

21.
Ripped skin. Ripping off tissue.

22.
A nose.

23.
Hole in the center of an ear.

24.
Exposing canal and bone

25.
Missing teeth. One lower jaw is.

26.
Gone. A set of lips.

27.
The chunk of a chin.

28.
And the shells. Shells from Verdun.

29.
Are still there.

30.
Unexploded ordnance. Sunk.

31.
Into dirt pockets. Like seeds.

32.
This blooming. Metal war.

33.
Shrapnel that looks like rocks or.

34.
Smooth egg of a bird.

35.
Soil made of mud and men and metal.

36.
How. Metal leaches and clings.

37.
This soil of war.

38.
Chlorine and lead and mercury and arsenic.

39.
Where every tree and every plant and every animal.

40.
Each blade of grass.

41.
Where 99% of everything died.

42.
Ground stripped raw.

43.
Stripped earth tissue or how this is.

44.
What war also.

45.
Also does.

46.
Damage to properties: 100%

47.
Damage to agriculture: 100%

48.
Impossible to clean.

49.
Human life impossible.

50.
The government declared it uninhabitable.

51.
A no-go zone.

52.
Broken skeletons of villages.

53.
And the craters that bombs make.

54.
Deep and round holes.

55.
How the bomb craters filled with water.

56.
Making. War ponds.

57.
This is a place.

58.
Where almost everything died.

59.
But the land.

60.
The land was still alive.

61.
Grass stretching again and.

62.
Grafting itself over the bone.

63.
Bone of what happened.

64.
Stretching over trenches and scars.

65.
Like new skin.

66.
And plants and trees and vines

67.
Rodents and snails and voles and mice.

68.
Deer. Wildcats with metal stomachs.

69.
Still living I say. To my husband.

70.
Who went to war.

71.
War that he did not want.

72.
Afghanistan.

73.
How he came home with hands and feet.

74.
Covered in blisters. Lesions the doctor said.

75.
Skin burning. Waking up to him crouched.

76.
On the floor and scratching. Saying I don't know.

77.
And I know.

78.
That this is how war is.

79.
Or later. I will lay in the darkness.

80.
And think about the burn pits in Iraq.

81.
Black smoke and jet fuel and fumes.

82.
About Vietnam sprayed. The bare mudflats after.

83.
Defoliation of trees. And birds. Missing mangroves.

84.
How dioxin poisons wind. Sleeps. In a river or sediment.

85.
The fatty tissue of a fish. Atomic blasts in Hiroshima and.

86.
Nagasaki. The incineration of bodies and land.

87.
Tearing skin off people. Tearing trees out of ground.

88.
Tearing everything.

89.
Away.

90.
How black rain fell. Radioactive bomb debris.

91.
Into mouths. Of people and rivers.

92.
How radiation lives. In grass and soil. The intestine of a cow.

93.
About the GWOT. Blood soaked years and streets and.

94.
How many miles of land. Where we let bombs.

95.
Unexploded or forever.

96.
I will think about Zone Rouge.

97.
Trenches like scars.

98.
My husband gardening. The tendons in his arms.

99.
Moving like trees.

100.
Or how war never goes away.

*"Zone Rouge" was first published in the *Wrath-Bearing Tree* on November 5, 2018. Reprinted with permission of the author.

Amalie Flynn is a poet and author of *Wife and War: The Memoir* (2013), *September Eleventh* (2021), and a collection of poetry blogs: *September Eleventh, Wife and War, The Sustainability of Us,* and *Border of Heartbreak.* Flynn's writing has appeared in the *New York Times, Time,* and the *Huffington Post* and has received mention from the *New York Times* and *CNN.* Flynn has a BA in English /Studio Arts, an MFA in creative writing, and a PhD in humanities. She is poetry editor for the *Wrath-Bearing Tree.* Flynn lives in Rhode Island with her husband and their two children.

Yoga and Animals: Inspiration for WWI Poetry

JANE SATTERFIELD

ELEGY WITH TRENCH ART AND ASANAS

The studio door swings shut and Emily instructs us
to begin in an *easy seated position.* Eyes closed, we begin

to *check in*, to *be mindful of each feeling that rises.* Time
to think of some *intention,* something as simple

as *the reason that brought you to your mat*—that, not the noise
rising up from Stoneleigh Lanes, the business at basement level—

thunderous, I think, though nothing next to the volley of
shellfire and mines going off in the front's busier sectors—

strange sound track stuck in my hearing long after
I left the museum's cool halls, since I walked the wood-

planked trench alleys of the Great War Exhibit. Overhead,
dangling on cables above us, the artifact highlights—Fokker,

Sopwith, and Albatross; a shiny Pfalz acquired as part of Allied
War reparations—great flying machines manned by pilots

in both strategic bombing and battlefield recon runs.
But I'm *here, now,* to take in and tune out the noise of summer

camp duckpin party cheers, the detritus that lingers
in modern memory from a war so literary, delousing

was known as *reading one's shirt...*[1] Trikonasana's next to
strengthen the core: your ribs drop into the body's

fuselage. A deeper bend into the archival absence of cordite,
the scent of rust that still rises, post-rainfall, from the soil

of stricken villages across the French countryside.
In almost any of the English market towns I visited as a child,

my grandfather said to take notice when we reached the town
center—each marked by a concrete cenotaph, memorial

to lost fathers, sons, uncles, brothers, and fiancés. You might
find a stray paper poppy lifted, then dropped by the wind.

Emily advises us to *find your edge.* Boredom was just one of
many unspeakably awful trench conditions—and the lull

between morning stand-to and evening assaults meant anyone
might be taken out by a sniper's precisely aimed bullet. The efficiency

of England's postal service meant letters arrived with stunning ferocity,[2]
making home seem distant yet paradoxically near. In situ,

soldiers chalked rocky outcrops; carved rings, crosses, or
pendants using spent bullet casings; lighters might be fashioned

from greatcoat buttons. In Tadasana, the yogi seeks strength
from the earth, each breath a means to recharge *so long as*

it's kept as an active position. The more elaborate work that's found
up for auction these days—a shell-casing vase with the image

of two wounded Tommies approaching Dover's white cliffs
with the word *Blighty* hammered out gently beneath—would have

been the handiwork of blacksmiths or engineers in rear areas,
valuable based on who signed it, when, and where. The lamp

my daughter noticed displayed in the *Believe It or Not* Odditorium,
(your "drop-by for a dose of weird")—was probably scrap,

surplus repurposed post-battle or ready-made by locals (one man's
meal ticket, another's battlefield souvenir). My illustrated history

tells me each war *is ironic because each one is worse than expected*,[3] though
this does not explain why we sustain, every few years at a time,

a renewable fashion for military chic. Dad was what? Ten, maybe
twelve, playing ball, when he lost the ring his father brought home,

something carved from the wing of a downed Messerschmitt.
How did his old man get it and where? When pressed

for more info on souvenir swag, when asked to decipher the image
code of badges or slogans on T-shirts from Gulf One squadrons

my father passed on to his sons, there's not one I remember.
My husband jokes that I'm either playing dumb or don't

want to recall. To face the fear that darkens each day,
I've been told I need to learn, really, to breathe.

This studio runs sessions guaranteed to get me *up off the ground*.
Slung in a hammock over a mat—that's one way

to target areas for maximum results. With an amazing
power to cleanse, this is something more than (though it

resembles) circus-based aerial art. Across generations, panic's
rarely made pretty—but what of those vases, hammered

from artillery shells, the designs almost totally obliterated
thanks to assiduous dusting? In Shavasana, we seek sensory

withdrawal, an entry into deeper meditative states. After,
we must rise again to collect our things, walk past

the salons and boutiques that line this block, walk back
into our separate worlds. Sometimes we stop in the bakery

to pick up a coffee, trying not to flinch at the giant screen TV
with its flashing, disastrous crawl.

BESTIARY FOR A CENTENARY

*Animals were used in World War I on a scale never before seen—and never
again repeated.*

–Alan Taylor, "World War I in Photos: Animals at War"

    1.
No one recruited them with posters
of trips abroad, obligations to protect
honor, or the "golden opportunity"
to "earn as you learn"—but
they also served in convoys &
cavalry, in pigeon schools & camel
      corps; on land, in sky & sea, as
      beasts of burden, scouts & spies,
      mascots & more.

    2.
Tales abound
of boon companions—
pack-mule #141's
unbudging stance midway
across a footbridge
signaled where a five-
nine shell lay ahead.
At signal corps' command,
dispatch dogs with harnessed
spools of wire crossed ground
socked by autumn storms
& mortars' unspeakable
rain, dodging craters
deep enough to swallow
men to laying lines for
point-to-point field telephones.

Others shouldered baskets
of carrier pigeons into
forward lines. Terriers,
shepherds, and bloodhounds
trained to search for wounded,
kitted out with Red Cross
gear in pouches on their backs,
heeled up with handlers
for inspection. Gas-masked,

outrunning sniper bullets, dogs
barreled through the lethal
fog of gas, that *Yellow Cross*
laid down by low-velocity shells
Doughboys & Tommies knew
as *Whizz-Bangs*, relaying
news through code notes
tucked in collars.

3.
Some admirable combination
of compass sense & mapping
genius made carrier pigeons
ideal couriers for conveying
communiqués & maps
of enemy terrain. When field
telephone lines went dead
under heavy fire, their homing
skills could be counted on

to get a message through.
Set free from baskets & carrier
boxes in the trenches, they
raised alarms from scuppered
ships, or planes droning
on a death spiral, zeroing
back to where they'd first
been set to roost, guided
by signature designs painted
on the roofs of mobile lofts parked
behind busier battle sectors.

4.
Of thousands employed
as urgent dispatchers in the age
before two-way radio broadcast
made mission tracking possible,
countless pigeons were intercepted,
shot, or shipped abroad, captives
paraded in home front streets.
Some were singled out for valor
& given medals with highest
honors. The last chance of a battalion
cut off in the Argonne, Cher Ami

flew through heavy fire, arriving
at his loft with a bullet in his breast,
message capsule dangling from
an injured leg. Today he stands
on one good leg, attentive as if
alive, poised in place among
entrenching tools, stick grenades,
prosthetic arms—museum memorabilia.[5]

5.
After battle rained down
grenades & gas in rounds
Yanks dubbed *Evening Hate* &
Enemy's Delight, one soldier's pet pit
bull, a stray smuggled from a shipyard
in the States, learned to track
no man's land for wounded men,
guiding the lost & disoriented
back to safety behind the line.
One day his bark alerted his unit:
an interloper sat trench side,
sketching, mapping out
the whereabouts of gunners,
ammo & supplies. Given rank,
"Sergeant" Stubby survived
seventeen battles & shrapnel
in his chest & leg, living on
to boost the morale of men
lodged in Red Cross recovery.

6.
The litany of what newfangled
combat wrought for horses
is a history itself; their cries
at night etched the air.

One veteran recalled that sound
troubled him more than cries
from fellow men: *Because we knew—well*
we presumably knew—what we were there for,
but them poor devils didn't, did they? No,
scores and scores of 'em fell to barbed wire,[6]
to bullets in the brain & worse
when mounts turned meals

for dogs & men. What to make of war's
strange economy: deep in combat zones,
glow worms, gathered in jars, gave off
reading light; even rafter rats crawled down
for crumbs of cheese, befriending
men in farmhouse billets *lying doggo*
before *zero hour* & Imperial sergeants
quipped that *silly blighters came for*
free while horses cost good money[7]

7.
A hundred years: in farmlands,
forests & cratered ground, munitions
& the bones of missing men
keep turning up; honorary mayors
serve the memory of villages wiped
off the map. We have no name
for the scale of pain. A photo
shows the redolent sadness
of a draft horse hitched to post
moments after its partner perished
from a round of shrapnel; podcasts
bring other accounts to life—standing,
waiting, a shield, a spot of shade
against desert sun, these & others
also served: comrades & battle kin.

AUTHOR'S NOTE

Like so many readers, I first encountered war poetry in the classroom—
through the lens of Wilfred Owen's "Dulce et Decorum Est." In its unflinching
gaze and brutal indictment of the gulf between public rhetoric and battlefield realities, it was completely different from anything I'd ever read. I'd go on to read
more WWI and WWII soldier-poets over the years, sometimes in class, sometimes
on my own—each encounter a powerful reminder of the way art provides a lens
of witness across generations.

I did not expect to contribute to the conversation as a poet. But in 1991,
on the advent of Operation Desert Storm, my father, an Air Force veteran and
reservist, was called up for active duty. He served as a quartermaster with the Congressional Airlift Wing. I watched—we all watched—tracers flare across the sky

announcing the start of battle and bringing seemingly distant conflicts directly into our living rooms. Over the next decade, as I read the work of veterans and civilians whose lives were touched by violent conflicts, I felt drawn to explore the legacy of war and the ways that inheritance resonates well beyond the battlefield into our homes, our language, and the daily fabric of our lives.

"Elegy with Trench Art and Asanas" began in a yoga studio with the yogi's invitation to be "mindful of each feeling that rises"[8]—an invocation that prompted the unexpected memory of a childhood visit to the Great War exhibit at the Smithsonian and my British grandfather's attentiveness to the WWI memorials in towns he took me to during a summer visit to his home in Corby, Northamptonshire, the town where my mother and I were born. In writing the poem, I allowed myself to "dive deeper" into personal and cultural memory, lingering over the artifacts that serve as touchstones and memorials for civilians as well as soldiers, in particular, an example of "trench art"—art assembled from wartime memorabilia, found objects, scrap metal, even bullets—that my daughter and I ran across at Ripley's "Believe It or Not" Museum in Baltimore.

In the summer of 2014, I ran across Alan Taylor's essay in *The Atlantic*, "World War I in Photos: Animals at War."[9] I was fascinated to learn about the many creatures pressed into service and, at the same time, how their presence provided the welcome companionship we experience with pets at home; this article was the inspiration for the sequence "Bestiary for a Centenary,"[10] which comprises the second section of my most recent book, *Apocalypse Mix*.

The practice of yoga—like poetry—is not a means of escape: it provides, instead, a space of healing, consolation, renewal; it deepens our connection to the world.

* "Elegy with Trench Art and Asanas" and "Bestiary for a Centenary" were first published in *Apocalypse Mix* (Autumn House Press, 2017). Reprinted with permission of the author.

Jane Satterfield has received awards in poetry from the NEA, *Bellingham Review*, Ledbury Poetry Festival, *Mslexia* and more. Her books of poetry are *Her Familiars, Assignation at Vanishing Point, Shepherdess with an Automatic*, and *Apocalypse Mix*, winner of the 2016 Autumn House Poetry Prize. *Daughters of Empire: A Memoir of a Year in Britain and Beyond* featured selections that received *Florida Review*'s Editors' Prize, the Faulkner Society/Pirate's Alley Essay Award, and the John Guyon Literary Nonfiction Prize. She is married to poet Ned Balbo and lives in Baltimore, where she is a professor of Writing at Loyola University Maryland.

Notes

[1] Paul Fussell, *The Great War and Modern Memory*: The Illustrated Edition (New York: Sterling Publishing 2009), 205.

[2] Fussell, *The Great War*, 201.

[3] Fussell, 7.

[4] Alan Taylor, "World War I in Photos: Animals at War," *The Atlantic* (April 27, 2014). https://www.theatlantic.com/photo/2014/04/world-war-i-in-photos-animals-at-war/507320/

[5] Smithsonian Institute. National Museum of American History. World War I Collection Sampler. "Cher Ami." https://www.si.edu/object/cher-ami%3Anmah_425415. Accessed July 2014.

[6] The Imperial War Museum. "Voices of the First World War: Animals in War." Accessed July 2014.

[7] The Imperial War Museum. "Voices."

[8] Jane Satterfield, "Elegy with Trench Art and Asanas," in *Apocalypse Mix* (Pittsburgh, PA: Autumn House Press, 2017), 6-9.

[9] Taylor, "World War I in Photos: Animals at War."

[10] Jane Satterfield, "Bestiary for a Centenary," in *Apocalypse Mix* (Pittsburgh, PA: Autumn House Press, 2017), pp. 27-34.

WWI Touches Pablo Picasso

DAVID ALLEN SULLIVAN

While poet David Allen Sullivan visited London's Tate Museum Great War Art exhibition in 2018, he was most struck by a painting that seemed almost irrelevant to the other artistic representations of battle carnage in the museum: "A Family by the Sea" by Pablo Picasso. To reflect upon his reaction, he wrote this poem:

The art of World War I,
displayed on the grey walls

of the Louvre's special exhibit
are numbing: ranked files
of death-masked soldiers,
faces crumpling into hands,
a brass general fashioned
from spent shell casing...

But in one eight-inch frame
beauty's hung—three figures
frozen in a pastel-hued beach scene.
I want to live inside this painting:
hovering shawl of Mother,
inquisitive pudge of Son whose finger
probes the still neck
of sleeping Father's sprawl.
                    *
The placard reads 1922,
says Pablo Picasso
was Spanish, so exempted
from France's war,
but not the war from him.
He was spit on
walking the streets of Paris—
too hearty, too well fed:
Qu'est-ce que tu fous lá?
(Why the fuck are you here?)
What did he say?
What could he have said?
                    *
He painted cubist distortions
of women, painted
his need for them,
painted them in pieces
while bodies burned,
were bayoneted,
gassed, gagged,
bound, bled, evicted diarrhea,
held up hands to have fingers shot off
so they'd be sent home—drank,
shat into holes they used petrol
to burn, grew gardens by trenches,
grew lonely, gave each other
hand jobs, haircuts, porn cards—
wrote letters, poems, prayers.

But Picasso's painting
is not about particulars—

its figures are statues
stolen from sun-baked Greece,
marbelized marionettes
on a suggestion of a beach.
Father's eyes are closed,
but the warmth of his nipples
stare at us, pale pink
of sea anemones, drifting—
their quiet hunger, their need.

*

Picasso never learned to swim,
feared Marie Therese
might drown so depicted her
being rescued again and again.
I imagine Marie loved water,
the cascade of it over her limbs,
the way her hands cupped
and pushed it back to pull her forward.
She both feared and loved him,
and contracted a disease
from the dirty waters of the Seine.

*

Picasso eases us into
this beach's simulacrum
of suffering. A dead body
washed up on sand—
or did someone drop it here?
Famille au bord de la mer,
is all his title tells us.
Blue grey sky layers down
to green metallic bullet sheen
of an ocean, then sand's band
of orange. The three stripes
form a dulled-out flag.
Mother rises up monumentally,
capstone of the three,
white wrap half undone, her hand rests
on the back of the Son. At the base
is the prone form of naked Father,
hand cupped over genitals.

The Son's arm, extends down
to press a finger
to the pulse-place of Father's neck.
More detail here than anywhere else
in the painting, death-mask
laid over his form.

The man's visage undoes
calm—Picasso's face
from the self-portraits.
                    *

Father by dull sea—
pieta of nobility—
no marks mar his body,
no contorting
from mustard gas inhalation.
What he suffered
has been drained from his visage.
The storm
that took him has moved on.
                    *

The dead walked
the streets of Paris
on crutches, on
improvised limbs, dead eyes
burrowed in, burrowed
through him, and Picasso
could't paint them into
or out of existence—
couldn't touch them.

Santa Cruz poet laureate **David Allen Sullivan's** books include: *Strong-Armed Angels, Every Seed of the Pomegranate,* a book of co-translation with Abbas Kadhim from the Arabic of Iraqi Adnan Al-Sayegh, *Bombs Have Not Breakfasted Yet,* and *Black Ice.* Most recently, he won the Mary Ballard Chapbook poetry prize for *Take Wing,* and published *Black Butterflies Over Baghdad* with Word Works Books. He teaches at Cabrillo College, where he edits the *Porter Gulch Review* with his students, lives in Santa Cruz with his family, and his website is: https://dasulliv1.wixsite.com/website-1.

How Do Wars Begin?

DONALD ANDERSON

How Do Wars Begin?
Today's newspaper: Wilfred Owen died on Armistice Day 1918, the last day of the war.

Owen's parents were informed of his death on Armistice Day. Owen had died the week before.

The newspaper is running the story in commemoration of the 100th anniversary of the assassination of the Archduke Franz Ferdinand of Austria—the act that starts the war.

Turns out that the assassin is one of nine children, six of whom had died in infancy. The newspaper reports a celebration in Sarajevo where the assassin, Princip, is treated as a hero.

In truth, it was Princip's accomplice who first tossed the bomb—a kind of grenade—at the motorcade. The bomb is deflected by the Archduke—hitting him in the arm. The Archduke is in an open car, so as to wave at his non-Austrian minions. The deflected bomb explodes late, missing the Archduke's car all together. Hard not to see it as a kind of cartoon: the Duke in a tallish hat with a lavender plume and the bomb a black ball with a visible fuse lit. Right? So here's the rest: the bomb thrower swallows a cyanide pill, then leaps into the nearby river. The cyanide pill, though, is expired, and the river that time of year is four inches deep.

The Duke—ever paternal—drives to the hospital to visit the injured victims of the car that actually caught the bomb. Big mistake, because Princip happens to be on the sidewalk when the Duke's driver takes a wrong turn. The Serb had been in a sandwich shop. While the driver is backing up, Princip shoots the Duke and then the wife when she tries to shield her royal hubby.

What if the Serb hadn't been hungry?

This is what starts the war: a mis-thrown bomb, a wrong turn in a car, bad cyanide, child non-mortality, lunchtime? Austria takes a hard line against Serbia and other powers in Europe choose sides. Within 30 days, a half-assed squabble between Austria and Serbia transforms into the first great modern war.

At the 100-year assassination site celebration, someone calculated that to hold a second of silence for every person killed in WWI in Europe that the attendees would have to stand mute for about two years.

And this: Princip's cyanide pill was expired too.

Donald Anderson's most recent books are *Quagmire: Personal Stories from Iraq and Afghanistan, Fragments of a Mortal Mind: a Nonfiction Novel,* and *Below Freezing: Elegy for a Melting Planet.* For thirty-three years, he was editor of *War, Literature & the Arts: An International Journal of the Humanities.*

How to Remember Your Ancestors' Names

DREW PHAM, veteran

I.
Your mother says her grandmother—
your mother's
father's
mother
had hair the color of blood spilled,
moon skin
a bridged nose
and no words
another missing term: người lai—
a person stitched
        together from
foreign bodies
you'd envy her features pale,
        were the cost
        not so
           dear
in her portrait she looks away,
        the same way
        mother would on
           her worst days
you wish you could read
        her the sutras
        as mother
once had
giving her letters
        kneaded into
a sentence to
 save you both
but her image tells you,
        nothing can undo
        seams sewn
           then torn

II.

  and what do they say to the little yellow men on the ramparts of verdun / im not a historian but a child-hood spent force fed movies about mass murder opens a window for me to imagine they tell these / soon to be / corpses to be brave but / not in a familiar tongue [hãy dũng cảm lên] instead i imagine those pale officers / say be brave / be / brave / be brave but / not the kind it takes to shrug colonial yokes from your shoulders / not enough to make a country from the spirit of an emancipating page / not enough to say no / no they lash these tirailleurs indochinois with their / naked european / tongues / baïonnette au canon / on ne passe pas / pour la france / a choice between steel / the bosches bullet and blade or the masters firing squad and guillotine / and one in three will die their mothers / names on their lips / the warmth of shroud-ed suns and flooded plains and star anise / a last mnemonic sensation before drowning in the absence of everything / but the rest / will wait their turns for one war to / metastasize into the next on and on but these little yellow / men will not be remembered / in the films pink / faced poilu kiss / rosaries and pray to saints bones enshrined in / spired crypts but no one says the prophets shahada / [لا إله إلا الله محمد رسول الله] / nor the heart sutra / [vượt qua, vượt qua, vượt qua bên kia, hoàn toàn vượt qua, tìm thấy giác ngộ] / and if they dont look / to god they stare / into lovers eyes clasped in lockets and wallets / held against / the breast but those so cherished / are too light / eyed and skinned / to look / like bà cố ngoại with her heirloom despair passed / down and down / to mother / to me and she tells me great⫽ grandmother was a pariah for reminding her / kith of the boots / on their necks the thieves / in their coffers the seeds / in their wombs / for that she'd been wed to another pariah / who'd taken up that colonial rifle and the thousands / like him sent ninety fold to dig their graves like veins / through the dysenteric earth / from belgique to suisse / but no one remembers / them the books i read / memorialize their spades but not their uniformed ranks though they died as napoleon said / for a piece of ribbon theyd hoped would make their people free / that invisible notion / liberté égalité fraternité / words to which they were / never entitled / no little yellow man could defend marianne in her phrygian cap / for such an act would sully that pristine ivory / flesh such an act reflects guilty / men's fear that what theyd done / to us we would revisit on them / when the great war ended those little men took their bits of ribbon home / braided those useless pretty strands into a hangmans rope / strangled an empire thatd been too quick to collect the blood / tax but loath to ever live up to their enlightened / promises / for that like a criminal washing gore from their hands the métropole erased their subjects bodies from the poppy sodden fields fed on indigène meat / but to take up arms against / an oppressor means / to take up his fire and powder and such wrath meets erasure with erasure / so i cannot know if ông cố ngoại had / bled over flanders field because / our self / appointed liberators remembered / the panoptic gaze of that blue white and / red banner / and chased all trace of that crippled empire from collective memory / chased the tens of thousands sent / to the western front / my family burnt every document and photograph and medal / to escape another oppression masquerading as reform / but itd been too late marking my mother / a felon across generations and i suffer no such brand yet i carry great-grandfathers crimes no less / i cant tell you which letters and tonals and syllables make up his name / i can only say / when i was old enough i also took up a rifle in the name of another falling empire / i did not know my inheritance / fighting for a cause / no one believes / yet it rattles through our throats without regard / lying in a hospital bed / mother beside / me she / was afraid she would never have the chance to make amends / told me this hidden detail / five more years before i saw his photograph / alongside bà cố ngoại's / mom doesnt like to talk / about the war or her parents war or her grandparents war / though it threads us together all / youd think that would close the gulfs and chasms between us but ive long since learned that common pain glues no / one together as much as i want to believe that i would understand the hollows and rises and caverns of their inner lives—i cannot—i only have this story / bits of shrapnel scattered through my family / pieced together but never whole / the explosion that tore its way through our roots detonated so long ago / i cannot tell you whether these bits of steel i still find in my limbs belong / to me / or the histories of my countrymen all so erased

III.
I spell my great-grandfather's name with
exposed skin and bowing hairs
beneath a mountain sun
my boot treads biting
strange soil are his
setting foot
on a country that called ours possession
and though I can't possess the memory
I see him
in terraced Afghan fields cut
up
    A
          hillock's
              sloped
              side
I see him in the farmers passed on patrol
they are him
shearing wheat from root / ankle deep gathering sickled rice
soldiers like me as common
as the meadowlark in spring fields foraging
and of theft,
I wonder if he ever killed a man and
if
that
were
true
I want to locate that slain soldier
in Western trenches
a yellow fist strangling a knife grip plunged between white ribs piercing white
skin made whiter with blood lost though I know
I'm projecting a past premeditation
because the gouverneur général left battalions behind
garrisoned not
to protect
but serve
    the

masters'
                                eternal
                                fear
                                        of
                                their
                  subjects
                  crescent
                  blades
      but
      I
know a nearer truth exists in between brothers
set one
against the other
I know what it means to take the master's lash as your own
because I have done the same
my body never lies
everytime I'm afraid
to look my mother in the eye
Great-grandfather, listen
your war's children have cut
our language from my throat
we have only tongues
shrapnel cleft
so hear instead tone and timbre
let sound express regret for
committed crimes
between shores both
foreign   and   domestic
but you should know
because my transgressions run so
deep I am alone
I imagine everyone
is a murderer who hasn't yet
had the chance
this isn't a confession, dear
grandfather
think instead
of the oily fog
from a stick of joss
our family's altar
where i write your memory
an offering unlike
fruit that rots
cigarettes that spoil

spirits made homeless
this offering remembers
when we are alone
for our trespasses
and when our names are
misplaced

<table>
<tr><td>

IV.
In this incarnation, I look at Great-grand-
mother
staring out from nearly a century away
and compare our faces in the mirror

when I was young
I wanted nothing
more than my face to blossom
from teenaged pores
into what I saw on TV
I insisted on getting limp-locked
haircuts that never worked
planned to change
my name to something
French like
Jean-Luc or Henri or Emmanuel
the highest compliment I ever received
from a girlfriend of one point one months
about my blood
—one part white
twenty-three parts Viet—

I always thought you looked European

when I was older and read more
I'd revised my desired name change to
honor
bà cố ngoại's father by
taking his—
Porc / Cochon
or whatever that filthy language calls swine

I know I am just a recent ripple in a pool
of blood
I am not possible without that oldest of
weapons
not the rifle
nor blade

</td><td>

V.
In an alternate reality, bà cố ngoại feels
no pain. When her mother tells her about
her father, she's wistful—it'd been love,
not rape, that'd joined the two, and there
is no colony there is no war that creates a
war that creates a war that creates my war.
There are no white skinned men who lord
over stores of rice they'd never sown nor
firebrands plotting liberation that isn't
needed nor empty bellies nor a million
famine dead graves. In this reality, my
Great-grandmother is not herself but a
woman who chooses, and Great-grandfa-
ther chooses, and they do not choose each
other, because they are not pariahs. In this
reality, she is not illiterate. In this reality,
she takes up the gifts I've ignored all my
life, her keen eye and a pen and sheaves of
a blank book, and I imagine she writes a
lyric for me—not as I do for her, in death,
but imagines me alive—and on that page I
flower from what would have been despair.
In this reality there are no secrets kept well
past their expiration, there are no histories
to uncover, there is no shame buried deep
within those histories. In this reality, she
takes her morning tea on the veranda of a
home that never existed in ours, listens to a
sea no refugee ever crossed, bites the drib-
bling yield of soil that's never been seeded
with bombs. She wonders who I might
have been, as I wonder her back to a life
never lived.

I stitch her body whole, the nots and
naughts—my thread, these keys—my
seamstress hands, this fantasy—my bolt of
cloth. I sew until my eyes are sore in the
dim light—my mother once told me to read
in the bright and sit up straight, and once,

</td></tr>
</table>

nor club
nor stray stone
the only weapon that at once makes and takes
life
so shouldn't I try to be generous?

but I compare
her to me
side by side
so little shared—
full lips and pale skin and a fold over the eyes
but her image my reflection
a sheet of glass a thumbnail of paper
an unmistakable resemblance.

that it was a shame I hadn't been born a girl, because, I think, she saw her grandmother in me, saw the need to teach, but her lesson had no plan,
just the tacit knowing that something passed between us each. I can't bring myself to end Great-Grandmother's life on this page. I can't built on opposites and absences. I can't reverse a
hundred years of hurt.

In this reality, she reads the sutra of the heart, translates a droning mantra for me—gone, gone, so far gone to that distant shore, oh! the other shore never left gone gone gone beyond utterly gone, hail this ecstatic emptiness! gate, gate, para gate, parasamgate bodhi svaha. In this reality, let me meet her on that empty shore, forsake life, for her sake—let me pay that price. Let me ease this stranger's way. Let her peer into the camera lens, see a future undone—here, we are birthless, deathless. Here, we are together in what never was, her letters tell me who I am, here, she has every word, so many beautiful words. Here, she tells me her name.

Drew Pham is a queer, transgender writer of Vietnamese heritage. A child of war refugees, her work centers on legacies of violence in times of conflict. She has published in *Blunderbuss Magazine, McSweeny's, Slice Magazine, Foreign Policy, Time Magazine,* the *Daily Beast, and Columbia Journal,* among others. She lives with her two cats in Brooklyn, New York, and serves as an adjunct English lecturer at CUNY Brooklyn College.

part ten

. WWI IN FICTION TODAY

A Distant Field:
America's Great War Highlanders

RJ MACDONALD, veteran

The mornings were cold in November 1917, so it's easy to imagine the scene: Her Majesty's Troopship Canada arrives at Liverpool docks, England. Soon lines of soldiers, burdened down with kit and rifles, are disembarking down the gangplanks. They form up into ranks, glad to be on dry land again, and with a nod from their commanding officer to the pipe major, the drone of bagpipes tuning up carries through the still air. Then, to the command of, "By the right, Quick March!" the men of the 236th MacLean Kilties of America march smartly away from the docks to the sound of their own pipes and drums. America's Great War Highlanders had arrived.

Americans serving in the Canadian Forces were commonplace during WWI. Three battalions, the 97th, 211th and 213th, all designated "American Legion," had been raised and deployed to France. But the 236th Battalion was special—it wore kilts. It was the brainchild of a Canadian Lieutenant Colonel Percy Guthrie. While lying wounded in France, he heard a Scottish battalion passing by, pipes and drums leading the way. At that moment he resolved to raise a battalion of Highlanders. On return home to Canada, he gained official support and in May 1916, the 236th New Brunswick Kilties (MacLean Highlanders) were raised. Needing men to fill its ranks, Guthrie's gaze quickly turned to New England following America's entry into the war. With a quick change of name and cap badges, the 236th Maclean Kilties of America appealed to Scots-Americans from Maine to Boston for recruits and within eleven days the Highland battalion had over-filled its ranks.[1]

Having arrived in Great Britain, the battalion trained hard, under the expert eyes of its officers and non-commissioned officers who were all veterans of the Western Front. In March 1918, they deployed to France. A cruel blow awaited them. Despite vehement protests, the battalion was effectively split into three to feed the constant need for reinforcements. The Canadians from New Brunswick were allowed to join the New Brunswick Regiment. The rest, including the Scots-Americans, were divided between the Royal Highlanders of Canada and the Seaforth Highlanders of Canada, in which they would go on to fight in every major battle for the remainder of the war, suffering 553 casualties.[2]

I came across the Maclean Highlanders while doing research for my WWI historical novel, *A Distant Field*. I was looking at Canadian Scottish regiments and stumbled upon a reference to the 236th MacLean Highlanders not only recruiting in New England but even changing their name and badges to the 236th MacLean Kilties of America. As my novel centered on Scots-Americans fighting with Scotland's Seaforth Highlanders, in which my great-grandfather served, I was immediately interested. I wanted to know more about them and where they served, with the possibility of having them pop up in future novels.

In 2006 a memorial plaque to the 236th Maclean Kilties of America was unveiled at their old headquarters building in Fredericton, New Brunswick. A memorial to all Americans who fought in the Canadian Forces during WWI was dedicated in 1927 at Arlington National Cemetery; the pipes and drums of the 48th Highlanders of Canada played at the ceremony.

One of the huge advantages of historical fiction is that you get to cherry-pick history. Instead of writing a comprehensive historical reference book covering a topic or period of time, you get to choose a few aspects and run with them. For me that became the American involvement in WWI prior to its declaration of war. Another great advantage is character. A character can reveal history through their persona-Americans volunteering or an Irishman signing up in the British Army by the thousands. One of my characters is a black Canadian, who only came about when I stumbled across a unit of Revolutionary War Black Loyalists resettled in Nova Scotia in 1783, which led me on to research black Canadian involvement in WWI. And then I've been able to mix characters with historical events. That meant for me, beginning with the sinking of the RMS Lusitania, a pivotal event not only in swaying American opinion behind the allies, but also a huge catalyst for American WWI volunteerism.

I also wanted to appeal to a wide audience, and historical fiction can do this if properly pitched. And therein lies the challenge: finding the right balance. Too many facts and the novel becomes a textbook. Too few, and it becomes fiction. I tried to find the middle ground by entertaining a good read, but at the same time leaving the reader with snippets of history along the way, all the while staying true to the military history of the time. I think of it as "teaching by stealth."

The other challenge is what to leave out. There is so much to cover when you're looking at a topic like WWI, but will it fit the storyline? Another danger is telling everyone what you know, or what you've learned. More than once I've had to remind myself, it's not a history book, it's a novel. The rewards for me comes when I discover an aspect like the MacLean Kilties, little histories that have become lost, but are waiting to be found again, like the Zion Mule Corps; the Royal Navy Trawler Service, in which another great-grandfather served; U.S. Cavalry on the Western Front; or the Maori Native Contingent-fearsome warrior soldiers from New Zealand who yelled their 'haka' war-cry as they attacked the Turkish positions.

Sometimes history and fiction co-operate. I was writing about four young Irishmen fishing in a rowing boat the day the Lusitania was torpedoed. I wanted them to be fishing for a reason, to sell their catch, so I checked what day the Lusitania was sunk—Friday. There's always demand for fish on a Friday in Catholic Ireland. A big smile. Looking ahead at novels down the line in the series, I needed the Seaforth Highlanders to serve alongside the American Expeditionary Force. In the Battle of Amiens, the British sent one crack infantry battalion to the American sector: Scotland's Highland Division. Another big smile. When events like that come together, it makes historical fiction so much easier for the author. The main events are known ahead of time—they're written in the history books, and the writer's job is to pick one or several then to weave a story around those events, to bring history to life, to make it real again.

Sometimes history and fiction don't cooperate. I desperately wanted my characters to visit the largely forgotten or unknown Salonika Front, but the dates and geography won't cooperate, so they're off to Iraq instead, to fight in WWI's desert campaign before inevitably heading to France.

RJ MacDonald grew up in Scotland and studied military history and social science at the University of California at Berkeley. A former U.S. Marine and Royal Air Force officer, his debut novel, *A Distant Field: A Novel of WWI*, won the 2019 Independent Press Award for military fiction, and a gold medal for historical fiction from the Military Writers Society of America who praised it by saying, "From the sinking of the Lusitania to the battlefield cauldron of Gallipoli, RJ MacDonald weaves an action-packed story that leaves the reader breathless." To see what he is currently writing, or to contact him, visit www.rjmacdonald.scot.

Notes

1 "The MacLean Kilties," Clan Maclean Atlantic, accessed January 21, 2022, http://www.clanmacleanatlantic.org/his-kilties.html.

2 Paul E. Belliveau, *Percy Guthrie and the Maclean Kilties* (LuLu Publishing Services, 2016).

Writing in the Post-War World of Agatha Christie

CHRISTOPHER HUANG, veteran

It's no secret that Agatha Christie's career as a mystery novelist began with the First World War. *The Mysterious Affair at Styles* was written at the close of the war, and Hercule Poirot made his debut as one of the numerous Belgian refugees seeking asylum in England during the war. Christie had served as a volunteer nurse, dispensing medications, and this experience becomes visible in her use of poisons in multiple books afterward. What is perhaps a little less obvious is the way in which the war's negative impact on society appears in her work. Part of this absence might have to do with familiarity; generally, there's no need to draw attention to something when an audience is intimately familiar with it. It might also reflect a distaste for unpleasantness; as a form of escapist literature, the whole point of the mystery story was to put the harsher realities of the post-war world aside and lose oneself in something a little more positive.

As someone writing more than a hundred years after the end of WWI, I do not have the same perspective that writers like Agatha Christie had of the post-war world. Its presence does not surround me or inform my environment. With hindsight and the distance of a hundred years, we're privy now to some objective, historical information that was unavailable to people living at the time, but the subjective experience must still be gleaned from the pages of then-contemporary literature. Agatha Christie may not have addressed the issues directly, but aided by what we know now, we can still get an idea of what it was like to live with the aftermath of the First World War.

Among Christie's works, the most obvious reference to the impact of the First World War must be *The ABC Murders,* which, early on, sets up a shell-shocked veteran, Alexander Cust, as the prime suspect in a spate of serial killings. She describes him as someone with an ineffectual personality, weak-willed, and prone to suggestion. We also know he'd received a head injury during the war and suffered from frequent headaches ever since. Because of difficulties stemming from his trauma, he's had difficulty holding down a job, though he has recently found employment selling stockings door-to-door.

His situation is not unique; one of the characters in the book comments on the persistence of an ex-Army man acting as salesmen—a qualification so specific as to suggest ubiquity—while another mentions feeling "sorry for these wretched men who go around and try to get orders." [1] Between them, they suggest the reality of war veterans living in reduced circumstances, engaged in an occupation that exists more as a form of charity than as a legitimate career option. The fact that *The ABC Murders* was published in 1936 also suggests that the eighteen years since the end of the war have not been sufficient for the hard-hit veterans to find their feet again. The off-handed brevity of the comments about veterans hints at a familiarity that readers in 1936 were expected to share and understand. These readers knew exactly why ex-Army men in particular might figure prominently as door-to-door salesmen, and why their entire category of men might be described as pitiful and wretched. The reality was that these WWI veteran-salesmen were alive, knocking right outside the front doors of ordinary readers at the time. *The ABC Murders* says more about their condition through its reticence than it could with any lengthy passage of descriptive prose.

A modern reader, however, doesn't have the same background historical knowledge of WWI, and a modern writer would be obliged to be explicit with the details.

Conversely, a modern reader does have a better awareness of PTSD than a reader of a hundred years ago. A few minutes of research provides a list of known effects ranging from the common to the rare, so we have some idea of what a traumatized veteran, such as we would find in the post-war world of the 1920s, might experience. A modern writer touching even briefly on WWI can be fairly certain of getting certain objective details right. However, even the experts were only just beginning to grapple with the phenomenon of PTSD in the 1920s and 1930s. As such, to understand the subjective attitudes of the general populace towards veterans with PTSD means turning once again to the literature of the period.

In *The ABC Murders*, a great deal of the tension hinges on the possibility that Cust might be subject to what we would today call a dissociative fugue state: that he might, in such a state, commit a string of murders without realizing it. In 1936, that suggestion made Cust easily credible as a suspect, perhaps more easily than it would today. But the actual medical accuracy or likelihood of a PTSD-induced fugue is unimportant compared to what the trope says about the popular perception of shell shock and PTSD. It reveals that people then were indeed afraid of a shell-shocked veteran's behavior and that they must have believed that the fugue state of Cust was a natural extension of war trauma. From there, the modern writer can extrapolate a certain wariness for those veterans with difficulty adjusting back to civilian life. It would be unsurprising as well to find that those same veterans would vehemently repudiate any suggestion that they've been adversely affected.

It is, I think, universal throughout the ages that no one wants to be thought mad if madness means being labeled a danger to one's surroundings.

Christie touched on these darker aspects of her post-war world in a light manner. However, the hints are there, and all it takes is some reading between the lines of history to form an idea of what lies just beyond the frame of the picture she paints with her novels. Writing now, a hundred years after the First World War, I find that getting the hard facts about historical events is easy; research will reveal how many people died, how many people suffered, how many people got away with barely any ill effects. For the subjective experience, however, literature provides the best clues, even when they are seemingly insignificant. As in Christie's *The ABC Murders* and as in her other works, the silences, the assumptions, and the inaccuracies can be as eloquent as the historical facts and figures.

Christopher Huang was born in Singapore. Following his mandatory two-year military service, he moved to Canada to pursue architecture. He now lives in Calgary, Canada, and writes full-time. His first novel, *A Genteman's Murder*, was published in 2018.

Notes

[1] Christie, Agatha. *The ABC Murders* (New York: Pocket Book edition, 1963), 150-151.

Wiring Party

ANDRIA WILLIAMS

This is an excerpt from my novel-in-progress, "Keeping." John Holm is a former British soldier who passed for white in WWI. Having immigrated to the United States, he falls in love with an Irish girl named Nessa. She disapproves of the war, and he is trying to think of something he can tell her to explain his feelings about it.

John felt hollowed and almost stunned from the dregs of his dream as he walked, faces passing him, all manner of haste or concern or sleepiness or ill health written in their expressions, and he wondered if any of them carried as much guilt as he did, if they had cause to, and how, if ever, they had unburdened it. He wanted Nessa there and yet, at the same time, did not; he felt untouchable and corrupt and also desperately lonely. The snow was matted gray at his feet, chunks of gritty ice in the gutter, yellow bulls-eyes of piss and dirty bladders of used condoms by the trash cans, two horses straining to pull a cart up ahead. John stopped to help push the cart from the snow. The driver thanked him for his help. John patted the horse nearest him, whispered hello. He was not remotely mystical but every horse seemed to have been someone or something else in another life; there was a gravity to them.

Helping the cart-driver made him feel just a tiny bit better, then better still as he walked and tried to see the faces around him with love, not anxiety, the way Walt Whitman maybe would, because Walt Whitman loved himself so much it made him love everybody else, too; and finally, after half an hour of walking, he thought that if Nessa were there he could maybe put his arms around her, he would love to sit with her awhile, a very long while in fact, sit for four days straight and hear her talk, hear her read that devilish Whitman that had hacked his heart open like a pike-pole on spring ice. He wanted to hear every tale of her youth and more, everything that made her happy, or as sad as he had just been, if there was anything that had, though he hoped not; and he wondered if she might listen to him too. He thought she would. He wondered, even if he told her the worst he'd ever done, would she forgive him. There was some magic in confessing to a woman, for if you did so and she loved you still, you were somehow absolved.

He'd dog-eared a page of the book the previous night:

This hour I tell things in confidence,

I might not tell everybody, but I will tell you.

As he drove north yet again, he tried to think of what he might tell Nessa that wouldn't make him feel ashamed, wouldn't make her doubt him, instead of the opposite. He realized how much he'd been worrying, how tense he felt. Each morning he'd awoken feeling as if his teeth had been sanded, his jaw tight and hurting. He tried to think of something he might like to write to her in a letter, something that would make her feel better about him, make him feel better about himself.

He could think of one time, on the Front. A wiring party. This was a story she might like. It was John and Brooks and one other bloke, whom John had only seen that one night and whose name he could not remember, and they'd been sent out to cut holes in the enemy's barbed wire. They'd blacked their faces with burnt cork so as not to shoot each other for Germans in the dark. Brooks had winked at John as if to say, Ain't that ironic, mate. John shushing him but finding it funny too. This was the winter before Arras, where Brooks had died.

John was snapping holes with as much haste as he could through the thick German steel, sweating beneath his helmet in the clammy drizzle, and he was embarrassed that he even cared that the others were faster; he felt inept with his dull cutters, each snip requiring a two-handed hold and gritted teeth and all manner of inward cursing at the absurdity of the thing. While he grimaced and snipped, he counted in his head the seconds between explosions along the line, a habit. Thirty seconds: the spatter of machine-gun fire somewhere on the other side; then the wail of German five-nines in the distance, a burst and crump as they exploded. He popped a coil of wire and restarted his count, this time got to eight in peace and quiet before he heard a second machine gun talking back to the first, common and territorial as the chatter of a squirrel.

And then the flare came, bright as creation, white perlen falling to the ground in fern-shaped arcs. It came from their own side, inexplicably, some genius turning on the house lights right overhead as if they were the three-piece band. Fuck all, said Brooksy, and the three of them froze.

Then John saw that the third man was staring at something. John followed his gaze and what he spotted there made him jerk in fright too: a white face staring back at them, spotlighted like they were, holding clippers in his hands and standing stock-still. Boche, maybe eight meters off. Poor fool was out repairing the holes in the wire John and his boys had chopped a few nights before.

And off to the side, near the fresh wire he must have lugged across the way for patching, John realized there were two other Germans, the lot of them staring, so they were equally matched and equally fucked, three and three.

John looked at the first German, the one closest to him, and willed the man not to speak. Please do not speak. Please don't. It was spooky, their language; John still heard it in his dreams, sibilant and choppy at the same time, like an incantation spelling your doom. But they all stood silent. They were all too surprised. And John, who was closest to the Germans, raised his hands to show he had naught but the wire clippers in them.

Maybe it was cowardice. Maybe it was just they were all out there doing chores, cutting the stupid wire, and in that instant to die for it seemed pathetic. But John with his hands raised at chest level began, slowly, to back away. Brooks looked over, eyes wide like to say, Are we really doing this?, and John gave him a nod.

If any one of them had panicked that would have been it.

The white rain light above them fizzled out and they were covered in darkness again, and John and his men began to move faster, backing away across No Man's Land. Holy Christ, let this be the way they'd come; it was too dark to tell for sure if they veered off the path they'd planned they could easily fall into one of the twenty-foot craters left by constant shelling. He hesitated, glancing over his shoulder. From the trench some good soul blinked a red light for them to follow; he hissed to his mates, and they turned and broke into a run, high-tailed it gasping back to the wire, banging against it to find their entry point. No sense in trying to be quiet now. Another flare was imminent, from the Germans or their own beloved idiots. The Germans had light-shell rockets they sent out on parachutes, could make midnight look like high noon. But somehow they were through and Brooks yelling he was stuck, John tearing him free with a rip of fabric and skin. They spotted the line of lime to guide them back through the craters toward their trench, softly iridescent and poisonous and welcome, and they ran for it as the sizzle of another flare went up in the distance, and they made it.

The first time he'd started to tell that story—in a pub near his mum's former flat in Liverpool, where he spent far too much time drinking in the weeks after he got home—he got to the part where Brooks spotted the Germans by the light of the flare, and a man had interrupted him with a thrilled guffaw, said Betcha cut them Huns down like wheat in August.

John had seen enough of falling bodies to know they did fall cleanly like scythed wheat. But sometimes they also toppled slowly, like a building under demolition, or sometimes they just kept running with patterns punched in their chests like game die, or sometimes they staggered like someone knifed in a bar fight, and there were times he felt the only goal he had left in life was to not have to see someone fall in any of those ways again. Why he loved this story of the midnight encounter, if loved was the right word for it, was that they'd all left upright and breathing, left to live another day. Suddenly, not killing had seemed the only wisdom that remained on earth, and so it was shocking to return to a world that only wanted to hear how he'd bravely killed. Because, yes, those Germans could have gone on to kill British boys. And that thought tormented John, kept him up some nights. Even though he knew it was just as likely that they'd died of illness, or become prisoners of war—still, he could not tell that story, because it was true that they might have killed British boys, and so he never tried to again.

But he thought that Nessa might understand. She was not soft-hearted, not exactly, but she was fair. She might like that story for the reason he did: that, for once, everyone in it had survived.

He felt more resolved than ever that he would leave off working for Titus McAvoy, and take Nessa and Aoife with him, and they would find a life of plain and honest peace. He'd never point a weapon at a man again. He'd never hurt a thing. He'd give up eating flesh. He'd—what was the word—meditate. He'd walk softly like a practitioner of some extreme eastern faith.

He'd made this resolution before, but now it felt serious. It felt like a promise of the soul. He'd like to see how far he could take the idea, scrape whatever momentum he could into a pile and blow on it, watch it glow and rise.

Andria Williams is the author of *The Longest Night* (Random House 2016), a Barnes and Noble "Discover" pick and Amazon's first-novel pick for January 2016. She is a fiction and nonfiction editor for the *Wrath-Bearing Tree* literary journal.

A Pretty Tame One

BENJAMIN SONNENBERG

The inside of the Liberty Truck stank of sweat. Thomas Croft Neibaur was pushed into the far back, just behind the radiator. He could hear the drivers' conversation over the stuttering engine:

"How many guys can we fit?"

"I dunno, about twenty-five."

Private Neibaur had counted around twice that number. As the truck filled with bodies and clamor about the Argonne, Neibaur withdrew his notebook and began to write:

Oct. 7th, 1918
Mrs. J. C. Neibaur
Sugar City

Dear Folks,

I received your most welcome letter a few days ago. Was very pleased to hear from you and to learn that you were all well as these few lines leave me at present. I am having a very good time over here chasing guns and cooties and I believe the cooties are the worst for you can hide from the guns but not from the cooties. I'll tell you folks this war sure is hard on a man and nobody but a real man could stand it. One night a man is sleeping in a good warm place and the next night he's sleeping in a shell hole with nothing to cover with but a cigarette paper.[1]

He stopped for a moment. Probably best to provide a little comfort. Lessen their worry.

Our outfit was in the St. Mihiel drive, also the second battle of the Marne and the big drive at Champagne. Of course, I came out all right but we lost a lot of men and we all had some mighty close calls. But with the help of God a few of us were saved. I sure am delighted at the success we are having. We're heading into a forest now, called the Argonne. The boys are a little nervous, mostly excited.[2]

The pencil dulled. Private Neibaur looked over the draft, decided it would do for now, and tore it from the notebook. He folded it and placed it inside his front pocket, not without some struggle. It was an itchy uniform, woolen and hardy for the cold and rain of which there was no short supply.

"Writin' home?" one of the men of his regiment asked. An Alabaman, like all the others.

"I suppose so."

"Don't tell too much!" another chimed. "Or they'll scrap it."

Private Neibaur laughed. "I'll be fine. You have a pencil? Mine's..."

He showed his nub and was given a fresh one.

"Perfect," he said and crossed out the last two sentences. "Can't be too careful."

"Loose lips," one man said.

Private Neibaur was an Idahoan. He was born two years before the end of the century, and in thick farm country. He had been in France for a few months now, and except for the war and all it entailed, he liked the rolling hills and little villages and pastures. It reminded Neibaur of home, although France was severely lacking in sugar beets and russet potatoes. It also lacked his family, one of the oldest in Idaho. As he had been told since he first touched soil, he was the son, grandson, and great-grandson of pioneers. Bravery ran through the entire Neibaur line. It began with his great-grandfather, Alexander Neibaur, the first Jewish convert to a new religion in the American heartland. It was not easy for Alexander Neibaur: after the death of Joseph Smith, Alexander and his wife spent much of their lives defending themselves and their faith.

A half-century later, Private Neibaur encountered similar trials. He had joined the Idaho National Guard shortly before Wilson decided to make the world safe for democracy. Much to Private Neibaur's chagrin, he was placed with the 167th Alabama Infantry Regiment upon disembarking in France. As he nervously wrote to his mother soon after, "The boys I am with now are from the south. Of course, they are good fellows and all that but still they have different ways that seem a bit funny to me. And of course, I get a little lonesome at times."[3] It wasn't that they were bad; they simply had a greater affinity for drink.

But Private Neibaur did as he had been raised to do: he shouldered his burden and steadied himself on his three rocks: family, nation, and God. As he trained at Sandpoint, Idaho, and began to feel the first twinges of homesickness, he assured his family of one thing: "I sure would like to be back home with you again but still I realize that I am serving my country in time of need, and I remember the words of Sir William Wallace: "God armeth the patriot."[4]

Private Neibaur slipped on the duckboard and fell into the mud. He tried to lift himself up, but his hands sank into the quagmire. As he struggled, two hands came from behind and pried him from the muck. He scrubbed the mud from his eyes and teeth and greeted the two Alabamans. They smiled and slapped him on the shoulder.

"We'd have left you, too, Tommy!" one said. "But we don't want to attract rats."

Neibaur grinned. No, not bad at all. He supposed they all had to joke from time to time to ward off the dumps. Self-medication and all that. He supposed it was also healthier than alcohol. He picked up the pace and made his way to the company's interior dugout. Their sergeant was to announce something and had been secretive about it, even though everyone knew they were to go over the top. An enormous push was building up behind the lines. For over a week, Neibaur heard the incessant whine of engines, the sloshy churning of mud, the creaking wheels, the aeroplanes, and, of course, the pat-pat-pat of the reserves, marching in formation. The 167th had been on the front trenches for about ten days now, and Neibaur prayed to God for an offensive, anything that could end the monotony. And the lice.

Passing through the communication trenches, Neibaur met with the others and entered the dugout. It was dim inside, as most of the lanterns had gone out. The sergeant was standing in the center of the room, beside his bed, holding a single piece of yellowed paper. He read over it, then again, then a third time, and yelled, "Company!"

The room fell silent as the words came from the sergeant, with great hesitation: Prepare to stand-to." The sergeant continued with the directives, but the point of it was something called the Kriemhilde. A fortress of some sort. A month of assaults spent hammering away at the walls. Soon, Private Neibaur stopped listening. Euphoria swept over him. He had fought before, certainly: at the Ourcq and at Soissons and the other unpronounceable names, but he'd never been part of a grand push before. He felt exhilarated. For the first time, Private Neibaur went up to the Alabamans and clapped them on the shoulder. "This is it!" he said, shaking them hard. "We're moving!"

"Tommy, you getting a rush of patriotism to the head?" one of them laughed.

They moved toward their positions and waited. Neibaur prepared his bayonet, cleaned the mud from his Lee Enfield, and for the first time in years, smoked a cigarette. When he was finished, he withdrew his notebook and a pencil, sharp this time. An hour passed as he mulled over what to write to his family. His rush of patriotism began to dull and was replaced with worry for the future. He supposed a few of the others had become desensitized, but he'd never been able to get over the chills that took him shortly before battle. But again, Neibaur steadied himself in the minutes before the sergeant blew his whistle that sent men over to machine gun nests. He repeated softly, "I am not going to die."

Then the whistle blew and Neibaur and the others hoisted themselves over the parapet. After ten days in a trench, the feeling of open ground was exhilarating and for a few seconds, Neibaur was back in euphoria. Then he saw a shell burst a few meters away, dousing him in a rain of blood and bits of flesh.

He stopped for a moment and then, watching the steam rise from what was left of the body, jumped into a nearby crater. The hole was filled with water and Private Neibaur fell up to his waist. Only a few years later did he realize how lucky he had been to jump into a relatively shallow crater. Men continued to rush past him, past the skeletal trees and into the fog that was collecting. Toward the Kriemhilde.

Neibaur pulled himself out of the crater and rushed forward into the fog. He began to pray again, but his words made little sense. He spewed forth promises to God if he would only let him live let him live let him live please please!

The fog stung his eyes. He closed them and tripped over a skeleton. He landed badly on a rock and felt an explosion of pain in his knee. Quickly, not sure if he had a minor bruise or a smashed kneecap, he lifted himself up and ignored the pain, though he could not keep from limping. His prayers became more fervent and less coherent. He talked to God about anything he could think of: his grandmother, crocheting by the fire. His mother, who would never receive the letter he wanted so desperately to write her. The sugar beets and farmlands.

He passed through the fog and only now began to realize that his battalion was almost gone. The rest of the men continued forward, out of the fog and skeleton forest and into open country. Neibaur squinted and, not too far in the distance, could make out a fortress seated on a modest hilltop. The other men had noticed it, too. An Alabaman whose voice he recognized shouted frantically: "There! There!"

Private Neibaur now saw why it had taken them a month to get this far. The Kriemhilde was a maze of deep, fortified trenches. Rows of barbed wire covered the entire slope of the hill that overlooked the area. At the top was a concrete bunker complete with gunports and shadowy slits. As he and the other approached, the black slits turned bright yellow as machine gun muzzles flashed and unleashed their loads.

He had never been this close to enemy machine guns before. He fell to the ground, hands over his head. He inched forward, afraid that if his rifle discharged or if he made a single noise, the machine guns would be on him. Neibaur made his way to the slope of the hill and hid under a tree. He saw the Alabaman inching toward him and called on him to hurry up. The soldier got onto his feet to make a run for it and his head exploded.

While Private Neibaur vomited, the orders came in in English and German. Were they really expected to use pistols, rifles, and grenades against entrenched machine guns? So, he did nothing and watched as waves of men, battalions reduced by half. Dozens of heads turned to powder. More men congregated around the slopes of the hill. They looked for leaders, who slowly came the leaders came for them. A sergeant would take control, and would be replaced by a surviving sergeant major, who would be replaced by a first lieutenant, who would be replaced by a single captain, and so on until no more captains emerged from the fog.

"All right, boys!" the captain said. "We're taking that bunker!"

A collective moan ensued.

"What?" the captain blustered. "Did you think we came all the way out here to enjoy the view?"

Quickly, prodded by incessant fire, they divided into groups of thirty. Neibaur huddled with his group. One of the soldiers looked to be no more than seventeen. He patted the boy on the back and gave him a cigarette. The other members of his group greeted him, each asking if Neibaur had a spare. Their accents branded them as Idahoans, probably from Caldwell or Meridian. Sure, he had spares. No rum, though.

After a moment's lull in the battle, the guns fired again, and planes flew overhead. The sergeant leading his group determined this to be a good sign. After another interminable wait, the captain blew his whistle and the hodge-podge battalion charged up the hill. The darkness turned the charge into a stuttering advance as Neibaur and the others struggled to find the enemy. The only way to progress, it seemed, was to move in the direction of the muzzle flashes.

Men were dying nearly every minute in the final push to the German frontline. Private Neibaur heard them groan and squeal in the darkness, and he prayed harder. Home. They wormed through the field, which now resembled the timber yard of his father's sawmill. Even the barbed wire the Germans placed on the perimeter resembled the kind his family would keep. He and the others, however, hated this wire, as it was the most dangerous point during any attack. It took about two minutes for a trained doughboy to withdraw his wire cutters and cleave through it. The Germans often attached cans to the wire, a sort of alarm that worked too well. The doughboy made it thirty seconds into his cutting before the shells or bullets found him.

Today, they were lucky. Or they had numbers on their side. They encountered no shells or bullets and Neibaur and the others made it. A trench stretched out a few feet in front of them, the very first of the Kriemhilde. The familiar sound of boots hit duckboard, and that guttural language. The young boy in Neibaur's group removed a grenade and threw it in. Neibaur followed suit and after an initial round of explosions, yelled at the others to hurry the hell up and get in.

Once inside, he reloaded his rifle. It was the first time he's been in a German trench. Stahlhelm were littered on the floor, along with the bodies of men killed by the grenades. The machine guns had to be moved, along with their ammunition.

The men continued down the trench until they came across a group of men near the end of the line. Neibaur's quavering finger was on his trigger and he was about to pull it when he heard one of the targets say, "Don't move, Kraut!"

"If you insist," Neibaur said.

"Well, shit," his would-be captor said. "Christ, let's go."

It was a long night, indeed. The Germans continued to pull back from their trenches, removing machine guns and shells. When they pulled back to the bunker atop the hill, they could withdraw no further and fought to the last man. It took grenades, mortars, and five hours to dislodge the Germans and finally conquer the Kriemhilde.

As the dazed prisoners marched down the hill and behind lines, Private Neibaur fell onto his knees. He had been spared again. But instead of praying, he removed his notebook and began to write to his family. He now knew exactly what he wanted to say, and how to say it. He was three paragraphs in when the captain he had seen earlier approached him and some of the others and asked if any of them would be willing to get a Medal of Honor.

"What're we doin'?" one asked.

The captain pointed westward to a small, grassy knoll, down the slopes of the hill and nestled between two copses.

"That's one of the last Kraut hold-outs," he said. "We've got some sharp-shooters giving us heck. We need three volunteers to help take them out."

Neibaur stood up.

"Good!" the captain said. "Who else is a man?"

The captain got his men together and gave them their orders. When the gilt of excitement melted away, nervousness took over. The knoll, thinly coated with wire, held two German machine gun teams. It simply had to be removed for the next wave of men to pass through, and the captain had decided—no doubt influenced by the events of the afternoon—that a small team was best suited for this operation.

"It's strictly up to you boys," he said. "You get this done, and we'll crack their Hindenburg Line. We'll be watching from here. When you've taken them out, signal to us and we'll do our bit. Good luck, boys."

Neibaur was designated the gunner and was summarily given the dreaded *Chauchat*. It was a French machine gun that Neibaur had heard referred to as the "Chau-shit." The magazine was of poor construction and its various openings meant that dust and grime and mud caked the insides of the weapon.

"Good luck, buddy," one of the volunteers said. "At least you've got your pistol."

They went out in the early hours of morning. Private Neibaur struggled to keep moving under the weight of the *Chauchat*. The team moved from rock to rock, boulder to boulder, bush to bush, and hole to hole. An obscene amount of time passed, but they finally made it to the ridge of the knoll. They drowned in sweat and panic.

Then their wire cutters were not working. One of the volunteers crawled to the entanglement and with all his might, pushed the cutters down onto the wire. In the darkness, Neibaur watched. No movement.

"What's wrong?" he whispered.

"Damn things won't cut through!"

"What?!"

"Shitting thing's dull!"

Neibaur and the other volunteer crawled to the wire. Each tried their cutters. In vain. It wasn't that their tools were spent; this wire was too strong. The entanglement consisted of many strings of wire bunched together. Too strong, no use, catastrophe. His mind boiled as he realized that they had no choice but to jump over it.

"Jump over it?" a volunteer whispered.

"We can't cut through and if we go back, we'll have delayed the division for hours. It's gotta be jumped."

They had ten minutes to prepare. Each had a bite of chocolate from their rations. The count down to three.

"Three . . . Two . . . "

One of the volunteers rose and leapt. Machine gun fire cracked and the entire knoll became bright.

"Go!" the remaining volunteer shouted and both men rushed. Private Neibaur threw his machine gun over the wire and leapt. His pants leg caught on the tip of the wire, and he tumbled. A pain in his leg made the earlier abrasion in his knee feel like sweet dreams. Landing on the grass, his hand shot to his mouth to muffle himself. Liquid was pouring down into his boots. Further up the knoll, the machine guns fired. A single hole had ripped into his pants.

His mind was screaming to leave but crawling back through the wire would be impossible. Neibaur looked for the other volunteers and ten feet away, he saw a rough outline of a body. He crawled closer to it, his leg shrieking with pain. The body was turned half on its side. Neibaur looked into the face and closed the dead man's eyes. He inched forward.

The firing stopped. Darkness swept the knoll and for the first time since the war began, Private Neibaur wept for ten minutes. He thought of William Wallace and of sugar beets, and pulled a line of gauze from his pack. He covered his leg as best he could and, feeling only slightly better, continued onward. He had only a little bit of ammunition himself; the rest was with the third volunteer's body.

It took Neibaur nearly thirty minutes to find the last body and ammunition crate. The man had been shot in the chest and had tumbled down the knoll. Neibaur went to him and closed his eyes, too. He pushed himself up using the *Chauchat* and removed the ammunition.

Another burst of light issued from the top of the knoll and the ground around Private Neibaur erupted. Dirt flew. Two bullets crash through his left leg.

Hi screams came, unmuffled. The firing stopped but he still grunted as he tried to ready the *Chauchat*.

Come on, come on, please don't jam.

He pulled the stands out and, fingering the magazine, loading every bullet. Neibaur pulled the operating handle back and readied the machine gun. Dozens of bodies seemed to rush down the knoll toward him. Bayonets glistened.

Neibaur pulled the trigger. The recoil was intense, and he lost his aim. He closed his eyes and moved the thing back and forth, side to side, hoping that when it ran out of bullets the Germans would all be gone.

The gun stopped with a Ping! Jammed again. His prayer had failed him. The Germans were still coming. Now they were close enough to begin shooting. The ground churned again. Heaving, he hit the side of the magazine and the Chaushit fired again. Now he kept his eyes open and focused on the Stahlhelm arrayed against him. Fewer and fewer remained by the second. Then a shot rang, this time a little closer, and the fourth bullet struck him in the hip. The world blurred and Neibaur let go of the *Chauchat*. He rolled down the slope. He thought of his mother.

Private Neibaur collided with a rock and his pistol fell from its holster. He reached for it and stopped. The thud-thud of boots on soft soil was becoming louder. It stopped beside him.

"Tot."

"Können sie da drauf achten."

Neibaur felt a boot lightly tap his rib. He made every effort not to cry out in pain.

"Tot."

They walked away. Neibaur waited. They were about twenty feet away, walking back up the hill. His pistol leaned against the rock. Neibaur grabbed it and steadied himself. The world came into focus. He drew a breath and fired again. And again. As the survivors retreated, he steadied himself on the rock and stood. Inches of his body worked against him as he waved his left hand. With his right, he fired his pistol into the air.

Men ran down the knoll again. Dawn broke as Neibaur added the last of his spare ammunition, cocked the pistol, and fired. With each man he felled, the charge slowed. He counted as they tumbled down toward him.

One . . . two . . . three . . . four . . .

Neibaur prepared to fire his last bullet when the counterattack stopped. The Germans had stalled. He called on them to throw down their arms: his command crossed the language barrier. As they held up their hands, he counted 11 eleven prisoners. Private Neibaur staggered to his feet. In the distance, a platoon of doughboys was coming to his aid. He did not wait for them. He brought his prisoners back to the American lines, a broad smile stronger than his pain.

Twelve days into his hospital stay, he began a new letter to his family. He ached everywhere and was burned except for his two hands, so he wrote to pass the time. He thought about telling them of his many surgeries and how the doctors had thought he would lose arms and legs, only to give him a miraculous bit of good news: four bullets would stay in his body forever, but he would lose no limbs. Instead, Private Neibaur focused on something else.

Thomas Neibaur Co.
167 U. S. Inf.
American E. F.
Oct 28, 1918

Dear Folks,
I have not had a line from home in a long time but then I know it is not because you have not written. I have been on the front for a long time then got wounded and am now in a hospital nursing a leg with three machine gun wounds in it. I am getting along fine the doctor says but it will be some time before I will be out again. Say folks I sure had quite an experience.[5]

He would indeed be getting some sort of medal out of all this, not that he had any need for such things. Still, they made quite a bit of fuss over him. An aged general with four stars on his shoulders came to him and shook his hand. He said he'd done about the most incredible thing a doughboy had done in the war yet. Then he assured him of the country's gratification and saluted Neibaur. Neibaur thanked him kindly, but asked if the general could speed up his hospital stay so he could get back to his family. The general said Neibaur needed his rest and left.

After the general, a man with a thick mustache came in to meet with Private Neibaur. He said he was writing a book on Medal of Honor recipients and wanted to ask a few questions.

"I'll tell you what I can, not that it's much."

"That's all I ask." He extended his hand. "James Hopper, pleasure to meet you."

Neibaur took it. "Likewise."

They discussed various things: Tom's family, his home-life, his training, his religion. Mr. Hopper seemed particularly interested in Neibaur's Mormonism.

"So, are you conflicted, Tom? Having to fight?"

Neibaur shrugged and smiled weakly. "Never gave it much thought, frankly. When you're getting shot at you don't think about much more than surviving."

The conversation turned to his regiment.

"What's the 167th like? You fight hard, huh?"

"I'll tell you how it was. They were so full of life and pep they had to be doing something all the time. But there were no boys who would stand by closer."

Mr. Hopper seemed impressed by this; he wrote in his notepad furiously. The conversation came to Neibaur's action on the 16th.

"What were you thinking? Why'd you volunteer, Tom?"

Neibaur smiled. "I don't know. Sudden rush of patriotism, I guess."

Finally, Mr. Hopper asked the last question of the interview.

"How would you describe yourself as a soldier, Tom?"

"A pretty tame one," he said.

Ben Sonnenberg interned with the United States WWI Centennial Commission in 2017. He graduated from the University of Maryland, College Park in 2018. His passions include writing, reading, and especially U.S. military history. He is currently a J.D. candidate at Harvard Law School pursuing a career in national security law. He thanks everyone at the Centennial Commission for giving him the honor of being featured in this collection.

Notes

[1] Thomas C Neibaur, "Thomas C. Neibaur to J.C. Neibaur, October 7, 1918," *Thomas C. Neibaur Papers,* ed. Samuel J. Passey (Rexburg, ID: Brigham Young University-Idaho, 2003), 1.

[2] Neibaur, October 7, 1918, 1.

[3] Thomas C Neibaur, "Thomas C. Neibaur to J.C. Neibaur, April 18 7, 1918," *Thomas C. Neibaur Papers,* ed. Samuel J. Passey (Rexburg, ID: Brigham Young University-Idaho, 2003), 1.

[4] Thomas C Neibaur, "Thomas C. Neibaur to J.C. Neibaur, June 2, 1918," *Thomas C. Neibaur Papers,* ed. Samuel J. Passey (Rexburg, ID: Brigham Young University-Idaho, 2003), 1.

[5] Thomas C Neibaur, "Thomas C. Neibaur to J.C. Neibaur, October 28, 1918," *Thomas C. Neibaur Papers,* ed. Samuel J. Passey (Rexburg, ID: Brigham Young University-Idaho, 2003), 1.

ACKNOWLEDGMENTS

I wish to first thank James Grieve for granting MilSpeak Books permission to use his photograph taken during the Pages of the Sea event created by director Danny Boyle for the 14-18 NOW Imperial War Museum initiative for the cover image for the book. I would next like to thank WWI Commissioner Monique Seefried for presenting this project to the WWI Centennial Commission and her unwavering support for me professionally and personally since 2016.

I also wish to express my appreciation to Mark Facknitz for sharing his knowledge and passion for WWI with me first as his student and teaching assistant and then as a colleague and dear friend. My understanding of war and trauma was further informed by important discussions with David Chrisinger, Connie Ruzich, Peter Molin, Seth Brady Tucker, Max Aue, Carine Trévisan, Anny Dayan-Roseman, and Cathy Caruth. Thank you to Rebecca Burnett for teaching me to use technological tools to foster projects in the humanities.

The blog would not have been possible without the assistance of Theo Mayer, the chief technologist and program manager for the WWI Centennial Commission, who offered help not only with the technical logistics but also with content and style management. My gratitude extends equally to Dan Dayton, executive director, and all the other members of the Official United States World War One Centennial Commission for including the WWrite Blog in their pedagogical mission.

To editors and board members of the *Wrath-Bearing Tree*—Andria Williams, Adrian Bonenberger, Ben Leroy, Michael Carson, Matthew Hefti, Amalie Flynn, Mary Doyle, Drew Pham, Rachel Kambury, Lauren Johnson, and Eric Robinson—thanks for the trust in making me one of yours and featuring my work.

Most of all, I am incredibly appreciative of Tracy Crow, Margaret MacInnis, Michelle Bradford, and their entire team at MilSpeak Foundation for their enthusiastic belief in this book, editorial guidance, and promotion efforts. To every writer who contributed to the blog and to this collection, I am grateful for your vision, talent, and, most importantly, for the time you dedicated to the project.

Finally, nothing would be possible without the support of my husband, Christophe; my daughter; my parents, Nancy and Robert (JJ) Orth; and all other family members—on both sides of the Atlantic. Thank you. *Merci*.

Photo courtesy of the author

ABOUT THE EDITOR

Jennifer A. Orth-Veillon, PhD, is a French-American writer, educator, and translator. From 2016-2019, she curated "The WWrite Blog: Exploring WWI's Influence on Contemporary Writing and Scholarship" for the Official United States World War One Centennial Commission. She holds a PhD in Comparative Literature from Emory University and led veteran writing workshops while completing a Marion L. Brittain postdoctoral fellowship at the Institute of Georgia Technology in Atlanta. A member of the board for the literary magazine, the *Wrath-Bearing Tree*, she currently teaches interdisciplinary humanities courses at the Catholic University of Lyon, the Lyon University Studies Abroad Consortium, and the Ecole Normale Superieure of Lyon. Her fiction, non-fiction, and translations have appeared in the *New York Times*, the *War Horse*, the *Wrath-Bearing Tree*, *Consequence Magazine*, *L'Esprit*, *Lunch Ticket*, and *Les cahiers du judaïsme*. Dedicated to exploring narratives of war through the arts, she has also written a novel based on the experience of her grandfather, a WWII battalion surgeon and concentration camp liberator.

Thank you for supporting the creative works of veterans and military family members by purchasing this book. If you enjoyed your reading experience, we're certain you'll enjoy these other great reads.

American Delphi
by M.C. ARMSTRONG

During America's summer of plague and protest, fifteen-year-old Zora Box worries her pesky younger brother is a psychopath for sneaking out at night to hang with their suspicious new neighbor, Buck London, who's old enough to be their father. Their father, a combat veteran, is dead—suicide. Or so everyone thinks, until Buck sets Zora and her brother Zach straight, revealing their father as the genius inventor of a truth-telling, future-altering device called American Delphi.

Salmon in the Seine
by NORRIS COMER

One moment 18-year-old Norris Comer is throwing his high school graduation cap in the air and setting off for Alaska to earn money, and the next he's comforting a wounded commercial fisherman who's desperate for the mercy of a rescue helicopter. From landlubber to deckhand, Comer's harrowing adventures at sea and during a solo search in the Denali backcountry for wolves provide a transformative bridge from adolescence to adulthood.

Cry of the Heart
by RLYNN JOHNSON

After law school, a group of women calling themselves the Alphas embark on diverse legal careers—Pauline joins the Army as a Judge Advocate. For twenty years, the Alphas gather for annual weekend retreats where the shenanigans and truth-telling will test and transform the bonds of sisterhood.

Collateral Damage
2nd edition
by KEVIN C. JONES

These stories live in the realworld psychedelics of warfare, poverty, love, hate, and just trying to get by. Jones's evocative language, the high stakes, and heartfelt characters create worlds of wonder and grace. The explosions, real and psychological, have a burning effect on the reader. Nothing here is easy, but so much is gained.

—ANTHONY SWOFFORD, author of
Jarhead: A Marine's Chronicle of the Gulf War and Other Battles

Sub Wife
by SAMANTHA OTTO BROWN

A Navy wife's account of life within the super-secret sector of the submarine community, and of the support among spouses who often wait and worry through long stretches of silence from loved ones who are deeply submerged.